I0112655

Soft Robotics

Intelligent Robotics and Autonomous Agents
Edited by Ronald C. Arkin

A complete list of the books in the Intelligent Robotics and Autonomous Agents series appears at the back of this book.

Soft Robotics

Cecilia Laschi

The MIT Press
Cambridge, Massachusetts
London, England

The MIT Press
Massachusetts Institute of Technology
77 Massachusetts Avenue, Cambridge, MA 02139
mitpress.mit.edu

© 2025 Massachusetts Institute of Technology

All rights reserved. No part of this book may be used to train artificial intelligence systems or reproduced in any form by any electronic or mechanical means (including photocopying, recording, or information storage and retrieval) without permission in writing from the publisher.

The MIT Press would like to thank the anonymous peer reviewers who provided comments on drafts of this book. The generous work of academic experts is essential for establishing the authority and quality of our publications. We acknowledge with gratitude the contributions of these otherwise uncredited readers.

This book was set in Times New Roman by Westchester Publishing Services. Printed and bound in the United States of America.

Library of Congress Cataloging-in-Publication Data

Names: Laschi, Cecilia author
Title: Soft robotics / Cecilia Laschi.
Description: Cambridge, Massachusetts : The MIT Press, [2025] | Series: Intelligent robotics and autonomous agents series | Includes bibliographical references and index.
Identifiers: LCCN 2024044334 (print) | LCCN 2024044335 (ebook) | ISBN 9780262049740 hardcover | ISBN 9780262383509 pdf | ISBN 9780262383516 epub
Subjects: LCSH: Soft robotics—Textbooks
Classification: LCC TJ211.44 .L37 2025 (print) | LCC TJ211.44 (ebook) | DDC 006.3071—dc23/eng/20250324
LC record available at https://lccn.loc.gov/2024044334
LC ebook record available at https://lccn.loc.gov/2024044335

10 9 8 7 6 5 4 3 2 1

EU Authorised Representative: Easy Access System Europe, Mustamäe tee 50, 10621 Tallinn, Estonia | Email: gpsr.requests@easproject.com

To Adriano
(finisce bene!)

Contents

     8.1   Overview   103
     8.2   Our Journey in Soft Robotics   103
     8.3   Soft Robot Abilities   104
     8.4   Soft Robot Applications   106
     8.5   A Vision for Future Soft Robots   107
     Self-Assessment Questions   107
     Further Readings   108

     Notes   109
     References   111
     Index   113

Preface

Soft robotics is a young field showing fast growth and transformative impact worldwide, which attracts the interest of research and industry. Despite being young and not yet structured as a discipline, it is becoming the subject of teaching in many universities, tutorials in scientific events, and training at different professional levels. A comprehensive view and knowledge systematization of this field is an endeavor that provides helpful, long-needed instruments for teaching. This book is conceived as a textbook to respond to this need for comprehensive and systematic teaching material in soft robotics. As a teacher myself, I spent time collecting appropriate materials for my own classes from the vast state of the art. More importantly, I made an effort to give the current knowledge a conceptual organization that can be applied generally and survive the fast-paced evolution of the field.

Soft robotics is an interdisciplinary field requiring diverse expertise. Teaching in this field also requires touching topics that traditionally belong to different educational pathways. My ambition with this book is to take a holistic view and include most of those topics related to soft robotics. After recalling robotics basics and describing the bioinspired approach that is often used in this field, the book covers the macro-areas of materials and technologies for actuation and sensing, modeling internal and external interactions, and soft robot control with model-based and learning-based approaches. The contents are organized in a modular way so that the book can be used as a whole, for a full course in soft robotics, or in parts, within other courses. As an example, the first chapter is an introduction to soft robotics that can be used at the beginning of a soft robotics course or as a soft robotics part inside another course (in robotics, for instance) or even as a self-standing seminar on soft robotics. The book modularity is helpful for teachers who may wish to teach only some aspects of soft robotics that fit their expertise. At the same time, the conceptual organization that this book is grounded in makes it easy for teachers to focus on specific topics and expand on them. The book modularity may also help in calibrating teaching for students at different levels and pursuing different majors, from undergraduate to graduate and PhD courses, from engineering and computer science to other disciplines related to soft robotics, such as material science, physics, and biology.

The different topics are treated as homogenously as possible, anchored to a common ideal case throughout the book. For each topic, the principles and theory are explained, as given by mathematical formulations; the technologies or techniques are described; and opportunities are offered to the students to challenge themselves in questions and further thinking. Supplementary materials accompany the book in a companion website, providing teachers with basics slides and videos and students with state-of-the-art examples and hands-on exercises. The website is intended to be a live repository with updates that help keep up with rapid progress in the field of soft robotics. It has the potential to become a live community for sharing teaching materials, descriptions of methods, exercises, and preprogrammed software for hands-on classes. Although the book is intended for teachers and students, it may also appeal to professionals interested in acquiring expertise in the field of soft robotics, which is still relatively new and not yet widely covered in the engineering curricula.

With the participation of readers and users of this book, soft robotics is ready to become a formal discipline and a topic of education.

Cecilia Laschi

Acknowledgments

First and foremost, let me acknowledge the excellent work done by this book illustrator, colleague and friend Irene Mannari. She has the rare ability to read my brain, generating clear graphics out of my foggy ideas and scrabbles. The book could not have been published without her professional contribution.

Everything that I wrote in this book is the result of years of interactions, scientific collaboration, and intellectual confrontation with many colleagues and friends. I have had the luck of receiving excellent mentoring and working with brilliant colleagues and students in my labs. I heartfully thank all those working with me in soft robotics at the BioRobotics Institute of Scuola Superiore Sant'Anna in Italy, at our Marine Lab in Livorno, where the journey started. And I thank the Soft Robotics Lab team members at the National University of Singapore for their enthusiasm and dedication to advancing the field of soft robotics.

Outside the lab doors, it is a privilege to be part of the small yet lively community that we have in soft robotics. We consider ourselves pioneers and inspire one another through elevated conversations in a friendly and creative atmosphere. Thanks to all of you. I am indebted to the Future and Emerging Technology (FET) programme of the European Commission for giving me credit to do the first soft robotics project (OCTOPUS) and for supporting the birth and growth of the soft robotics community through the Coordination Action RoboSoft. I am grateful to the Robotics and Automation Society of the Institute of Electrical and Electronics Engineers (IEEE) for giving us the opportunity to gather around the IEEE Technical Committee on Soft Robotics and for technically sponsoring our IEEE RoboSoft conference.

I wish to especially thank a few people who invaluably went through some of the chapters. Thanks for taking the time to read the material and for providing your critical views and technical comments. I wish to thank Matteo Cianchetti for reading chapter 4 and verifying the technologies described there with his knowledgeability and well-known accuracy. I also thank you, Matteo, for reviewing the history of soft robotics and its milestones in chapter 1. Without Federico Renda's opinion on internal interaction modeling, I would not have dared to publish chapter 5 on modeling. Thank you, Federico, for overseeing the content and organization, and for your patience in checking all equations. For reviewing external interaction modeling, I thank Gianmarco Mengaldo who, as my colleague and co-author, brought me closer to computational modeling. Soft robot control is a slippery topic, difficult to conceptualize and present systematically. I thank Egidio Falotico for reviewing chapter 6 and helping me maintain focus on presenting general concepts, without slipping into the myriad specific cases. Thanks, Egidio, for your critical comments and accurate clarifications.

On a personal note, I wish to thank my (large) family for always making me feel supported. I am grateful to my parents for being role models. I also thank my late father for catching our first octopus and, last but not least, my mother for cooking for me in the last days before book submission.

Cecilia Laschi

1 Introduction to Soft Robotics

Chapter Objectives

- To understand the rationale behind soft robotics
- To define soft robotics
- To identify the main milestones in the field
- To learn about the technologies and techniques involved in soft robotics

1.1 Overview

We are about to start a journey in soft robotics. This chapter is bringing you to the start line by first convincing you that soft robotics is worthwhile. We are going to understand the motivations for this scientific and technological endeavor, and we will retrace the steps of the first pioneers. Definitions are cornerstones in learning a new field, and we start our journey from there. As always, it is not an easy task to give a comprehensive definition of a field, but we will at least build our terminology. We will try to give a face to our ideal soft robots so as to anchor our learning points, using examples. The challenges that pioneers had to face stand as the building blocks of the description of this field. Their achievements are the milestones that guide our journey. Let us start.

1.2 Why Soft Robotics?

A journey inside soft robotics cannot but start from robotics. What is robotics in fact? What can be considered to be a robot? In what ways is a soft robot still a robot, and how is it different? Starting with robotics, we must say that it is a very special discipline, young yet solid. Roboticists consider 1960 as the birthdate of robotics,[1] when the first robot, Unimate by Unimation Inc., was installed in a General Motors plant in the United States. It is not a long history for a scientific or technological discipline. Since then, roboticists have built a vast body of knowledge on how to design, build, model, and control robots, and manufacturing companies have vastly employed robots in their processes. In terms of accuracy, repeatability, and overall reliability, robotics technology has performance that is hardly matched by other technological fields. Today, robots are employed in a variety of fields, from manufacturing to medicine, from space to the abyss, from household to personal assistance. A diversity of robots exists, in terms of shape, technologies, and applications. Just think of the robot manipulators used in factories or the futuristic, legged humanoid robots. Still, what is a robot? This question continues to be a challenging one and an opening for most robotics textbooks.

A generic robot has sensors, actuators, and some level of intelligence in order to plan its behavior (i.e., for executing actions in response to sensory inputs).[2] What makes robots different from digital devices is their *embodiment*: a robot is a physical machine, immersed in a physical world, which it perceives and acts on. The definition given by Maja Matarić in her 2007 book synthesizes from previous definitions and captures the essence of being a robot.

Definition 1   Robot   A robot is an autonomous system that exists in the physical world, can sense its environment, and can act on it to achieve some goals. (Matarić 2007)

Embodiment emerged as a key requirement in the 1980s, as it appeared unrealistic to aim at an intelligent machine without providing it with a body in addition to an artificial brain. "Intelligence requires a body," in the words of Rodney Brooks, one of the fathers of robotics, at MIT.

Definition 2   Embodiment   The robots have bodies and experience the world directly— their actions are part of a dynamic with the world and have immediate feedback on their own sensations. (Brooks 1999)

A step forward in this direction is the concept of *embodied intelligence*. In the study of intelligence, it was becoming more and more acknowledged that the body does not just have a role in exploring the environment; it also has a role in shaping intelligence itself. Intelligence is better intended here as sensory-motor behavior, relevant in robotics. Rolf Pfeifer, a scientist in robotics and artificial intelligence at the University of Zurich in Switzerland, brought this idea forward. According to the embodied intelligence hypothesis, sensory-motor behavior is not completely planned by the brain and shaped by the nervous system. A part of it comes from the mechanical adaptation of our body to interaction forces. Our movements, though controlled by our brain, are partly mechanical, passive movements. As an example, when we walk, our knees compensate for terrain roughness with no brain involvement; it is the joint compliance that makes our legs adapt to the external reaction forces from the ground. Similarly, when we grasp an object, our fingers adapt themselves around its shape because of joint compliance and soft tissue. So, interaction forces guide the final movement, still following the grasping motor command coming from the brain.

Definition 3   Embodied intelligence   The computational approach to the design and understanding of intelligent behavior in embodied and situated agents through the consideration of the strict coupling between the agent and its environment (situatedness), mediated by the constraints of the agent's own body, perceptual and motor system, and brain (embodiment). (Cangelosi et al. 2015)

The physical body contributes the part of intelligence and behavior that emerges from the complex interaction with the environment. As such, embodied intelligence depends heavily on the body mechanical properties, its shape (or morphology) and the arrangement of perceptual, motor, and processing units).[3] We need soft, compliant bodies to take advantage of all this.

1.3 What Is Soft Robotics?

Just take a moment and think of a robot that you may know from the news about the achievements of a fancy research lab or from your favorite science fiction movie. Then take a look at your pet, playing around you in the house while you are reading this book. What is the first difference that comes to your mind? Robots tend to be rigid, in body and movements, while animals look soft, even though they have rigid bones inside their body. Robots can be more accurate, in fact. They can do precision tasks in factories and be very fast. Is it because of an error in evolution that we are not as rigid? From the previous section, you can argue that this is purposeful and necessary for using the embodied intelligence inside our bodies. In fact, we still exceed robots in some other tasks, such as when we quickly adapt our movements to the environment around us.

It is now clear that soft robotics is about introducing compliance in robot bodyware. How do we introduce such desirable compliance in robot bodies? We can act on the way robot movements are controlled. With proper sensors and mathematical calculations, we can make the robot adapt to external forces using so-called interaction control techniques. But we can even act on the way robots are built, by using soft materials or deformable structures. This second pathway is the topic of this book.

A simple definition of soft robots is that they are "robots built with soft materials" (Laschi and Cianchetti 2014). Most definitions focus on the materials used for building soft robots. A more specific definition considers their Young's modulus, a measure of material "softness," and describes soft robots as "primarily composed of materials with moduli in the range of that of soft biological materials" (Rus and Tolley 2015). Still focusing on materials, "soft-matter robotics" as used in the literature is based on the concept of "soft matter," well-known in material science (Wang and Iida 2015). Since we observe that the vast majority of animals are soft-bodied, "soft-bodied robots" is also an analogy, a term used to compare robots to "soft-bodied animals." This term is not only intended for animals that are composed of soft tissues alone, like worms or octopuses. These are quite special cases, in fact; these animals are either very small or live underground or underwater at high environmental pressures and with altered gravity. What is more interesting is that even animals with stiff skeletons are mainly composed of soft tissues and liquids. To give an example, the human skeleton accounts for a scarce 11 percent of our body mass (Kim, Laschi, and Trimmer 2013). However, we should enlarge our view to the overall compliance of the soft robot, regardless of the materials it is made of. One of the first definitions points out that "a soft robot is inherently compliant and exhibits large strains in normal operation" (Trivedi et al. 2008). In 2014, a group of soft roboticists gathered with the specific purpose of creating an international scientific community of soft robotics. They realized the importance of a definition as a fundamental step to pursue their purpose. The following definition is from this RoboSoft community:[4]

Definition 4 **Soft robotics**   Soft robots/devices that can actively interact with the environment and can undergo "large" deformations relying on inherent or structural compliance.

This definition is the most comprehensive because it outlines the deformability of the robot bodyware, either by intrinsic material compliance (i.e., low Young's modulus) or by extrinsic morphology that magnifies strains of rigid-material structural elements (i.e., plastic or metallic thin layers or fibers). It is time for a few examples to help visualize the concepts that we have touched on. Let us have a look at figure 1.1.

1.4   Brief History of Soft Robotics

The term "soft robotics" flourished in the 2010s, yet robots corresponding to our definition of a soft robot appeared much earlier. We can refer to them as *ante-litteram* soft robots, to emphasize that they were just named differently.

Already in the 1950s, after serving in the Manhattan Project, Joseph Laws McKibben invented a type of pneumatic actuator intended to approximate human muscles. They are based on a pneumatic chamber inside a braided-fiber structure. When inflated, the McKibben actuators can contract, which is why they are associated with muscles. In fact, they are used in couples to reproduce the agonistic-antagonistic arrangement of most animal muscular systems. His invention was inspired by polio, a widespread disease in those years that affected thousands of children worldwide, including McKibben's daughter, depriving her of her upper limb functionality. Her father aimed at building a robot arm as an artificial limb that could be soft and safe but powerful at the same time, which is also important in soft robotics. McKibben actuators are also still very relevant, and you will learn all the details of their working principle and technology in chapter 4.

As you see, soft robotics is very much rooted in the attempts to reproduce principles observed in living beings. This methodology is referred to as *biomimetics* and, when applied to robot design, as *bioinspired robotics*. The idea is straightforward enough to go on reading this chapter, but you can learn more about the methodology in chapter 3. Biomimetics drove the development of soft robots in the 1990s and beyond, drawing inspiration from a wide variety of animal models. For one example, researchers in Northern California overcame the dislike of cockroaches to create robots inspired by them that capture the essence of biomimetics, soft robotics, and embodied intelligence. Cockroaches are excellent runners and can quickly adapt to different terrains without missing a step of their fast run. In robotics terms, it means adjusting a high number of degrees of freedoms, corresponding to all the joints that the animals have in their six legs. A pretty complex computational task. However, cockroaches are very fast and do not need to do much computing in their limited nervous system. The compliance of their joints helps them. Professors Mark Cutkosky, a roboticist at Stanford University, and Robert Full, a biologist at the University of California at Berkley, distilled the principle for fast, adaptable running in cockroaches. They developed a series of fast-running robots, providing them with compliant joints in their legs. Their robots use mechanical feedback from the terrain to adapt their steps, with no software calculations. This is how they use their embodied intelligence. Despite not being named as such, these robotic cockroaches deserve to be considered as ante-litteram soft robots.

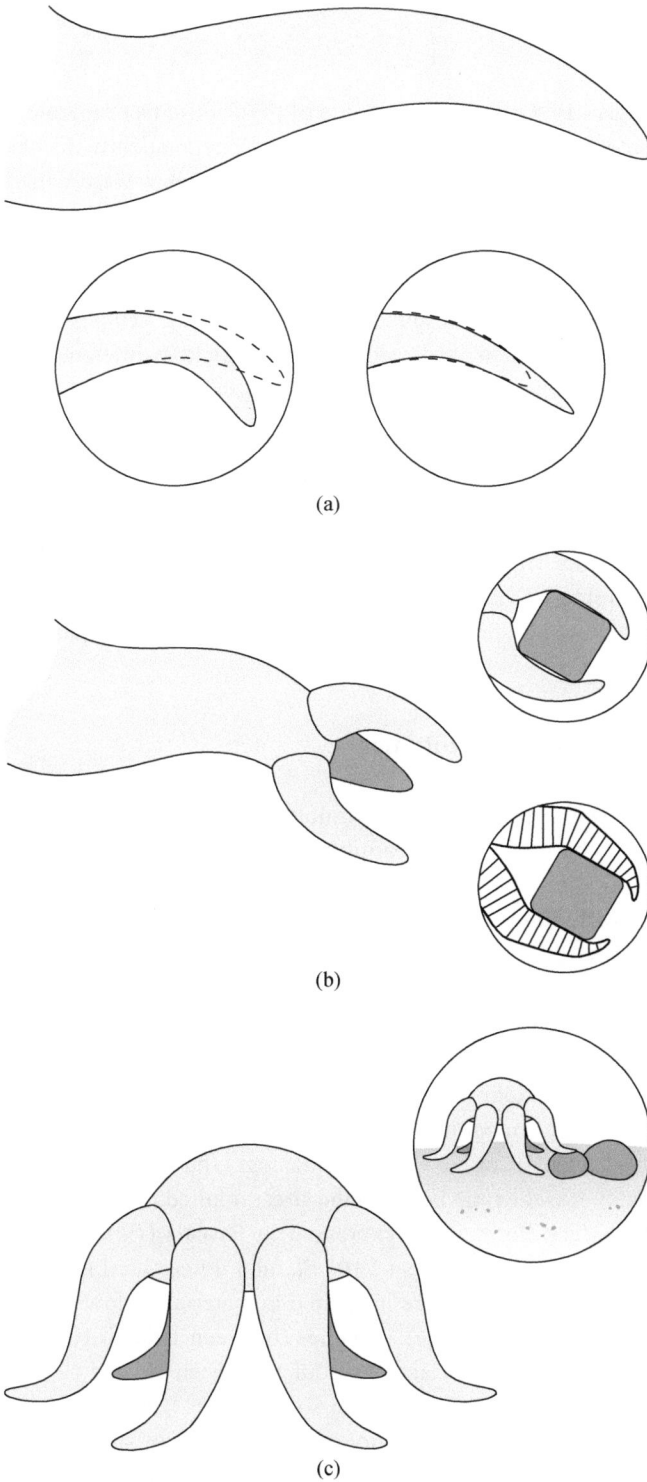

Figure 1.1
Conceptual representation of soft robots. For the purposes of this book, let us focus on (a) rod-like soft robot arms, (b) fingered grippers, and (c) legged robots, well aware that one of the key advantages of soft robots is the variety of morphology they can have. Rod-like arms in (a) can bend but also elongate. See how the fingers in (b) are soft either for their materials (right-top) or for the geometry of their structure (right-bottom). In the latter instance, the material is rigid, not soft per se. The soft legs in (c) adapt themselves to the terrain, increasing walking effectiveness and facilitating its control.

At the same time, still in the early 1990s, on the other side of the Pacific Ocean, Koichi Suzumori was testing the flexible microactuator (FMA). Actuated pneumatically, FMAs consisted of fiber-reinforced rubber chambers and were proposed for what were called "gentle miniature robots with no conventional links."

At the end of the first decade of the 2000s, the term "soft robotics" was being used explicitly, according to the definitions given in section 1.3. The first examples of soft robots in the following years were still bioinspired, like those inspired by elephant trunks, octopus arms, and caterpillars. A few initiatives gave a boost to the development of the field in different parts of the world—for example, the US Defense Advanced Research Projects Agency (DARPA) program on ChemBots, the Chinese program Tri-Co Robot, and the European Integrating Project OCTOPUS, followed by the Coordination Action RoboSoft. In the next decade, most research funding agencies launched purposive soft robotics programs. International scientific societies supported the growth of the community, especially the IEEE Robotics and Automation Society, and a purposive conference was started in 2018, IEEE RoboSoft.

The field of soft robotics is extremely lively and productive, and the soft robotics technologies developed so far represent a solid body of knowledge that you are just about to discover in this book.

1.5   Challenges and Milestones in Soft Robotics

When drawing the history of a discipline, the exact chronological order of events is not always what interests one most. What matters for learning lessons are insights into the challenges faced during the evolution of the field and the corresponding achievements that represent the milestones of its growth. Figure 1.2 summarizes them. Let us take this viewpoint, asking ourselves the questions that tormented soft robotics pioneers.

1.5.1   What Materials Are Needed for Soft Robotics?

Starting from our definition of soft robotics, we need either a soft material or a deformable structure to build a soft robot. In this section, we focus on the first case. "Soft materials" is not an accurate definition. The Young's modulus of a material, or modulus of elasticity, is a common measure of a material's tensile or compressive stiffness (chapter 4 recalls the mechanical properties of materials). It is the ratio between the stress applied to the material and the corresponding strain (i.e., its deformation), expressed in Pascals (Pa). To give some references, hydrogels have a Young's modulus of 10^4 Pa, and a very hard material like diamond has 10^{12} Pa (see also figure 4.4). We are interested in materials below 10^9 Pa ($1 GPa$), like some rubbers (between 10^7 and 10^8 Pa), silicones (between 10^5 and 10^7 Pa), or even very soft materials like hydrogels, with a Young's modulus between 10^4 and 10^5 Pa. In practice, silicones are the most popular materials used in soft robotics because of their easy handling and versatility. Since silicones can be molded in a variety of shapes, soft robot prototypes can be easily built and tested in research laboratories. More interesting materials can provide soft robots with additional abilities, like self-healing. For instance, some hydrogels can recover their original mechanical properties after being cut if the two parts are put in contact and some trigger is activated, like heat or light.

However, this is just the start of the endeavor. We need materials that move in a controllable and programmable way.

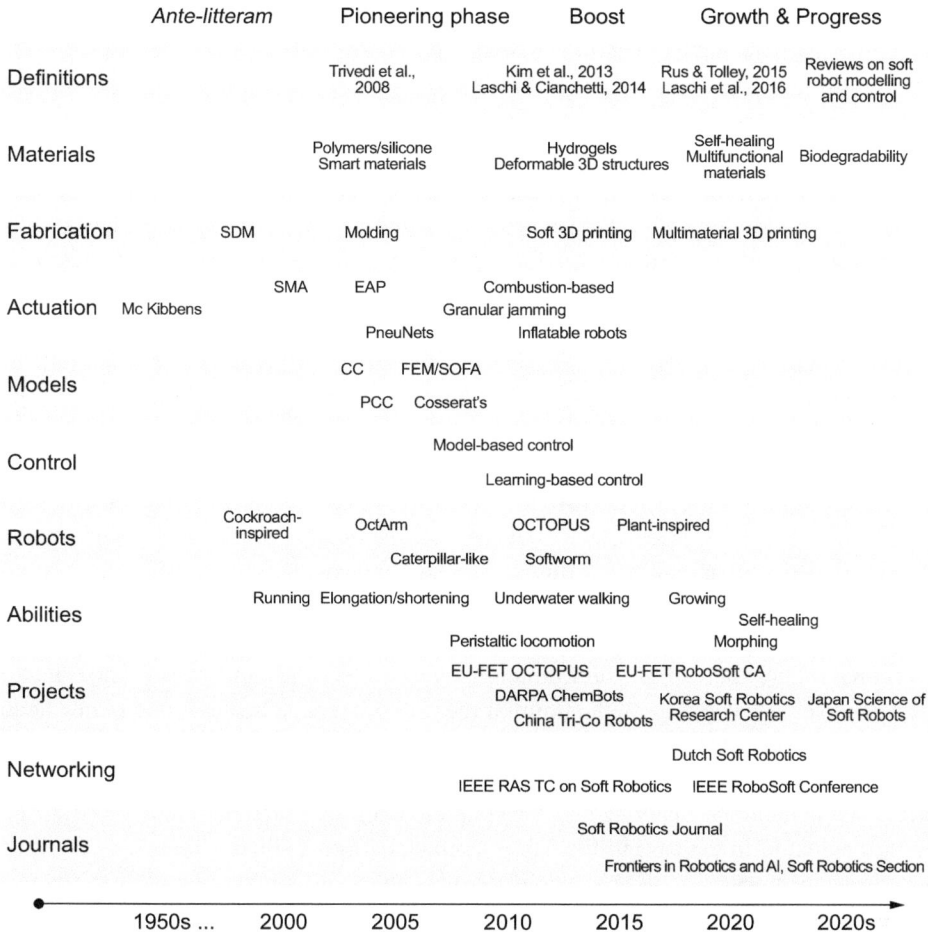

	Ante-litteram	Pioneering phase	Boost	Growth & Progress
Definitions		Trivedi et al., 2008	Kim et al., 2013; Laschi & Cianchetti, 2014	Rus & Tolley, 2015; Laschi et al., 2016; Reviews on soft robot modelling and control
Materials		Polymers/silicone; Smart materials	Hydrogels; Deformable 3D structures	Self-healing Multifunctional materials; Biodegradability
Fabrication	SDM	Molding	Soft 3D printing	Multimaterial 3D printing
Actuation	Mc Kibbens	SMA; EAP; PneuNets	Combustion-based; Granular jamming; Inflatable robots	
Models		CC; FEM/SOFA; PCC; Cosserat's		
Control			Model-based control; Learning-based control	
Robots	Cockroach-inspired	OctArm; Caterpillar-like	OCTOPUS; Softworm; Plant-inspired	
Abilities		Running; Elongation/shortening; Peristaltic locomotion	Underwater walking; Growing; Self-healing; Morphing	
Projects			EU-FET OCTOPUS; DARPA ChemBots; China Tri-Co Robots	EU-FET RoboSoft CA; Korea Soft Robotics Research Center; Japan Science of Soft Robots
Networking			IEEE RAS TC on Soft Robotics	Dutch Soft Robotics; IEEE RoboSoft Conference
Journals			Soft Robotics Journal	Frontiers in Robotics and AI, Soft Robotics Section

1950s ...   2000   2005   2010   2015   2020   2020s

Figure 1.2
An illustration of the history of soft robotics and its main milestones.

1.5.2 How to Make a Piece of Soft Material Move

One of the first challenges in soft robotics is coupling a soft material with some kind of actuator. A very simple first step is to cast silicone in the desired shape and to embed tendons (e.g., inextensible cables) inside it. When a cable is pulled from outside, the material is deformed. For instance, in a rod-like arm, a lateral cable generates a bending by shortening one side of the arm. Figure 1.3(a) provides a visualization of this mechanism. An antagonistic cable, or just the material elasticity, can bring it back to the initial shape. This is perhaps the simplest form of a soft robot and is still very effective. The cables can be pulled by electrical motors, located outside of the robot, and the movements can be controlled by software controlling the motors.

Instead of pulling the soft material on one side, we can obtain a conceptually similar deformation (i.e., a rod-like arm bending) by stretching the material on the other side. What if we mold an empty chamber inside the soft robot body and inflate it with a fluid? The answer is shown in figure 1.3(b). Let us take again our rod-like arm, create a longitudinal chamber on one side, and fill it with a fluid. When the pressure of the fluid is increased, the chamber tends to inflate. The arm is stretched on one side and tends to bend to the other side. In fact, you do not really want a balloon-like inflation, and a variety of ideas have been implemented to constrain the deformation and obtain the desired movement, as you will learn in chapter 4. Actuators based on this approach are called flexible fluidic actuators (FFAs), and they are very popular because of their versatility and effectiveness. Designing the morphology of FFA-based robots, setting the shape of chambers, and adding the constraints to obtain the desired deformation is a creative process that gave rise to a variety of fluidic soft robots.

However, some materials can move, or change shape, after some form of trigger. They are called "smart" materials, to emphasize their responsiveness to external stimuli, or "kinetic" materials, to stress their movement ability. To some extent, they can serve the purpose of soft roboticists by giving motion to a soft piece of matter. In this big family, electro-active polymers (EAPs), and their dielectric varieties (DEAs), have long been known as "artificial muscles" because of their capability to contract. Thanks to the Maxwell stress effect, two electrodes separated by a soft medium tend to attract each other when a voltage is applied to their endings. A film of silicone with conductive opposite surfaces can easily achieve this effect; see figure 1.3(c).

Another big class of smart materials are shape-memory materials. They can recover their original shape, after a plastic deformation, by an increase of temperature. The original shape can be set as desired through an initial thermal process. When deformed, they keep memory of the original shape and, when heated, go back to it. The material itself is generally a metallic alloy, so it is not really suitable for building a robot body out of it. In practice, shape-memory material can be coupled with soft materials and serve as an embedded actuator. In the case of the cables embedded into a silicone body, as described above, shape-memory alloy (SMA) wires can play the same tendon-like role. Instead of using motors to pull the tendons, their shortening is given by the recovery of an initial, shorter shape. This can be obtained by heating the SMA wires by controlling a flow of electrical current through them. A spring-like morphology amplifies this contraction, as depicted in figure 1.3(d).

Chapter 4 gives the details of the actuation mechanisms mentioned here, and summarized in figure 1.3, and describes the underlying working principles and the mathematical relations involved as a set of tools for designing them.

1.5.3   How to Model a Soft Body

In robotics, obtaining a mathematical description of a robot is at the base of controlling its motion. We need a geometrical description of its mechanical structure and a model of its kinematics (i.e., the movements it can make). We also need a model of its dynamics—that is, how it behaves when actually moving, considering the forces that generate this motion and the effects of its body dynamics, like inertia and friction. Chapter 2 recalls these robotics basics. The techniques for robot kinematics and dynamics modeling are well-known. However, they assume that a robot is a chain of rigid links connected by joints that allow relative motion of two consecutive links, which is not really what a soft robot is.

From what we learned in section 1.2 about the definition of embodied intelligence and in section 1.3 about what a soft robot is, we can depict a model of a soft robot as a deformable body, with internal forces produced by its actuators and external forces coming from

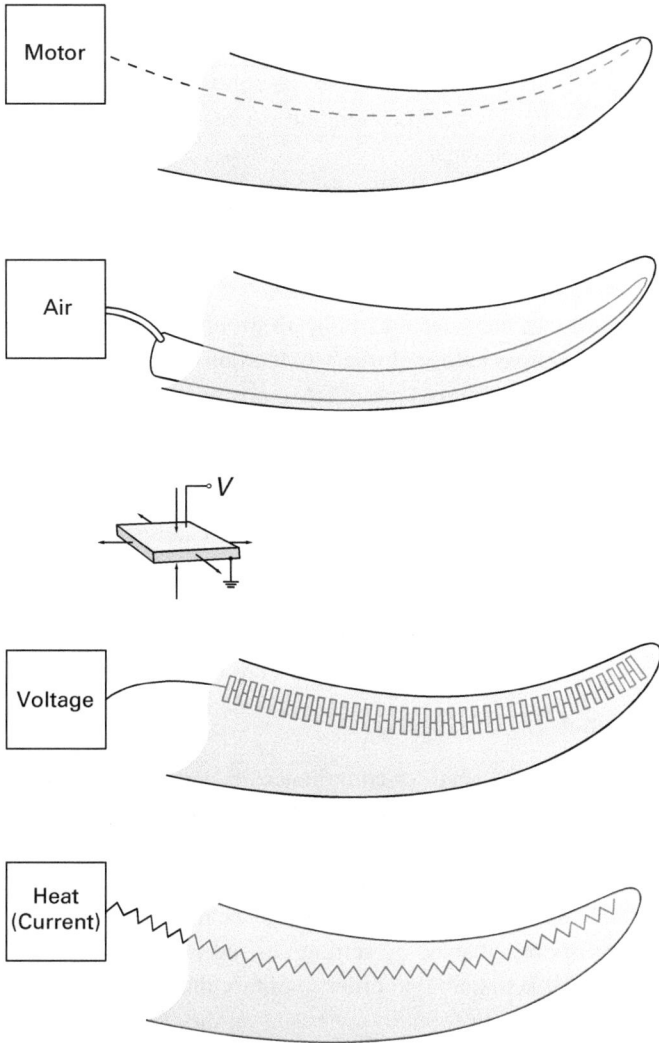

Figure 1.3
Conceptual examples of a soft arm bending obtained with (a) a cable on one side pulled by an external motor;
(b) a fluidic chamber on one side, pressurized with an external air supply; (c) EAP stacked on one side; (d) an
SMA wire or spring acting like a cable on one side.

interaction with the environment. We learned in section 1.2 that the soft robot behavior is
given not only by actuation but by a combination of such internal forces with external
forces. Modeling them also gives quantitative inputs to design. Chapter 5 teaches how to
model internal and external interactions for control and design purposes. To what extent
can we adopt the same approaches and techniques used for rigid-link robots? Let us use
our rod-like robot from figure 1.1(a) as an example and see a few possible approaches.

The well-known finite element method (FEM) is used in soft robotics to capture thor-
oughly the deformations of our soft arm. Broadly speaking, the soft robot is modeled
through a finite number of points that are connected to form a mesh, approximating its
3D body and its deformations. When the soft body is rod-like, as in our example—meaning
that one dimension (length) is much larger than the others—so-called rod models can be

employed, and deformations can be described by the position and orientation of the points of its backbone. A general model used with rod-like soft robots is the Cosserat's model, which assumes the rod as a continuous set of infinitesimal micro-solids stacked along the backbone. The deformations are described by the position and orientation of frames set on those backbone points, thus including torsions.

The methods above refer to continuum mechanics. We may wish to describe the behavior of the soft body through finite-dimensional parameters instead. When possible, we approximate our robot as a curved link, which can be described by a set of arc parameters. We may assume the curvature constant, with a *constant-curvature* (CC) approximation. If the soft robot is not uniform in shape, we consider it as built up from a finite number of curved links, each of which can be approximate with the CC method. This approach is called *piecewise constant-curvature* (PCC) approximation. A more general approach is considering a finite set of points on the backbone of our soft robot arm and setting reference frames on them. We describe deformations through their rotations and translations with respect to each other.

Similarly, for external interactions, we may adopt approaches in the realm of continuum mechanics or finite parameters. Both cases of interaction with fluids and with solids need to be considered with both approaches. Today, they are also complemented by data-driven methods that can find modeling relations through learning.

If you cannot wait to learn how to use all these methods, jump to chapter 5.

1.5.4 How to Control a Soft Robot

We said that robots are soft to respond to the need for compliance, as necessary to implement embodied intelligence. Further, we said that embodied intelligence uses external interaction forces and body deformability to simplify the computational tasks involved in control. So, a part of the motor behavior of our soft robot is just obtained mechanically. We learn more about this in chapter 3. However, we may wish to control our soft robot in the same way as we control any robot—that is, by setting desired positions in space and calculating the inputs to give to the actuators—in order to obtain the required movements to reach our targets. Typical methods in robotics are based on models of the robot to control, or they make use of learning techniques. The same two kinds of approaches can be used in soft robotics, with some differences in practice.

Model-based soft robot control
Robot models for control purposes consist of relations between the poses (positions and orientations) of the robot end effector, expressed in the joint space and the Cartesian space, respectively. Those relations allow one to transform a position in joint space into a pose in Cartesian space, and vice versa. Especially, the latter inverse transformations are used to find the joint configurations corresponding to the desired robot pose. Since actuators usually act directly on the joints, this is needed to make the robot move as intended. In soft robots, we do not have a similar relation between actuators, joints, and robot pose. The concept of joints is not always appropriate for soft robots, and the end effector is not always relevant to the task. What we consider as a general case in soft robots is that actuators deform the robot, which corresponds to a displacement of its tip in Cartesian space. We have seen before (section 1.5.3) that we have tools for modeling soft robots and finding a mathematical

description of the relations between actuators and deformation. We can use them for controlling our soft robot position in space. Similar to the case of rigid-link robots, we can invert the relations and find the inputs for the actuators in order to reach the desired position. Chapter 6 explains how to implement model-based control techniques for soft robots.

Learning-based soft robot control

What if the relations discussed in section 1.5.3 are learned by the robot itself? Machine learning techniques are evolving rapidly and have quite a long history of application in robotics. In robot control, they serve exactly the purpose of building the relations between joint and Cartesian space. Neural networks are often used in this case and many other learning tasks in robotics. The control systems based on neural networks are often named *neurocontrollers*. Although the type of neural networks adopted can vary greatly, in general terms, we can say that the learning phase consists of robot movements where a neural network creates the mapping between the two spaces. Such movements can be done on a robot simulator, but most often they are done on the physical robot. This learning phase on the physical robot is extremely relevant in soft robotics to encode the deformations that are complex to model and to take into account the physical characteristics of the robot.

1.6 What's Next in Soft Robotics?

Before we continue, a fair question to ask about soft robotics is how long this current knowledge will last. As a young discipline, soft robotics is still growing and evolving. Investigations are underway all over the world, and new achievements occur often.

Soft robotics research is still focused on the basic ingredients: actuation and sensing technologies or modeling and control techniques. More and more, however, soft roboticists are starting to explore the applications of soft robots in practical tasks. They include operations in the biomedical field, monitoring of natural environments on land and sea, as well as some industrial tasks in sectors that require more versatility in the manipulation of objects, like the food industry. Chapter 8 presents a few examples.

The progress of soft robotics is also driven by progress in material science. In this sense, soft robotics is progressing toward robots that can self-heal, grow, biodegrade, and embed sensing and motion in the material they are built of. Likewise, progress in modeling and control is constantly contributing to building and using new soft robots.

Keeping pace with the continuous production of new knowledge in soft robotics is simply not possible in a book. It is not even a well-set purpose. The soft robotics community is very prolific, and the rich production of scientific papers gives excellent insights into new findings in this field. This book is focused on the basics that the discipline is built on, standing as the cornerstones for its understanding in its present and future forms.

1.7 Summary

We have entered the field of soft robotics by giving definitions and a bit of history, limited to little more than a couple of decades. In fact, we have given proper acknowledgment to ante-litteram testimonials, since some technologies and theories emerged much earlier. We have seen how soft robotics is motivated by the observation of the role of soft tissues in living beings and by their embodied intelligence. You have learned the main challenges that soft roboticists have been facing, in a variety of different scientific and technological fields. Addressing them brought a number of important technological milestones that are shaping the field. We acknowledge that we cannot keep pace with the vivid and rapid progress in this field, but we are going to learn what constitutes its fundamentals in the rest of this book.

Self-Assessment Questions

1. Describe an example of soft robot.

2. A braided sleeve is equipped with a cable anchored to its tip and pulled out on one side of the base. Is it a soft robot?

3. Which soft robotics milestone do you find more impactful in the field?

4. Which challenges faced by pioneer soft roboticists do you find more engaging?

5. What learning paradigms do you find more suitable for controlling soft robots?

Further Readings

On Soft Robot Modeling and Control

Calisti, Marcello, Giacomo Picardi, and Cecilia Laschi. 2017. "Fundamentals of Soft Robot Locomotion." *Journal of the Royal Society Interface* 14 (130): 20170101. https://doi.org/10.1098/rsif.2017.0101.

Della Santina, Cosimo, Christian Duriez, and Daniela Rus. 2023. "Model-Based Control of Soft Robots: A Survey of the State of the Art and Open Challenges." *IEEE Control Systems* 43 (3): 30–65. https://doi.org/10.1109/MCS .2023.3253419.

George Thuruthel, Thomas, Yasmin Ansari, Egidio Falotico, and Cecilia Laschi. 2018. "Control Strategies for Soft Robotic Manipulators: A Survey." *Soft Robotics* 5 (2): 149–163. https://doi.org/10.1089/soro.2017.0007.

Mengaldo, Gianmarco, Federico Renda, Steven L. Brunton, Moritz Bächer, Marcello Calisti, Christian Duriez, Gregory S. Chirikjian, and Cecilia Laschi. 2022. "A Concise Guide to Modelling the Physics of Embodied Intelligence in Soft Robotics." *Nature Reviews Physics* 4: 595–610. https://doi.org/10.1038/s42254-022-00481-z.

On Soft Robotics and Embodied Intelligence

Laschi, Cecilia, Barbara Mazzolai, and Matteo Cianchetti. 2016. "Soft Robotics: Technologies and Systems Pushing the Boundaries of Robot Abilities." *Science Robotics* 1 (1): eaah3690. https://doi.org/10.1126/scirobotics .aah3690.

Pfeifer, Rolf, Josh Bongard, and Simon Grand. 2007. *How the Body Shapes the Way We Think: A New View of Intelligence*. Cambridge, Mass: MIT Press.

On Soft Robotics Technologies and Materials

El-Atab, Nazek, Rishabh B. Mishra, Fhad Al-Modaf, Lana Joharji, Aljohara A. Alsharif, Haneen Alamoudi, Marlon Diaz, Nadeem Qaiser, and Muhammad Mustafa Hussain. 2020. "Soft Actuators for Soft Robotic Applications: A Review." *Advanced Intelligent Systems* 2 (10): 2000128. https://doi.org/10.1002/aisy.202000128.

Lee, Chiwon, Myungjoon Kim, Yoon Jae Kim, Nhayoung Hong, Seungwan Ryu, H. Jin Kim, and Sungwan Kim. 2017. "Soft Robot Review." *International Journal of Control, Automation and Systems* 15 (1): 3–15. https://doi .org/10.1007/s12555-016-0462-3.

McEvoy, Michael Andy, and Nikolaus Correll. 2015. "Materials That Couple Sensing, Actuation, Computation, and Communication." *Science* 347 (6228): 1261689. https://doi.org/10.1126/science.1261689.

Pons, José L. 2005. *Emerging Actuator Technologies: A Micromechatronic Approach*. Chichester, UK: John Wiley & Sons. https://doi.org/10.1002/0470091991.

2 Robotics Basics

Chapter Objectives

· To recall a few robotics basics—namely, robot mechanics, sensing, and control—that are needed in the next chapters

· To build up a conceptual scheme of a robot as a guideline throughout the book

2.1 Overview

In chapter 1, we agreed on a definition of a robot.[1] We also observed how diverse robots can be, in their morphology, functions, abilities, and application fields, still responding to the same definition. In this chapter, we leverage the conceptual elements of the definition to give the fundamental components of robotics that fit any robot. We are not focusing on soft robots here. We are going to use a conceptual scheme of a robot, as a guideline throughout the chapter, accompanied by examples that can help fix the lessons learned. We are going to build this conceptual scheme step by step, by adding robot components. If you wish, you can jump ahead to figure 2.17. It gives us the final picture, and you may choose to start there instead.

According to this strategy, we are going to learn about the classical robot mechanisms and how their movements are controlled. We are going to learn how forward and inverse kinematics and dynamics are used in robot control and what the classical control systems are, for purposes of controlling robots in their task space or in joint space. To do so, we need to know more about typical robot sensors and actuators. We are interested in both proprioceptive and exteroceptive sensors (i.e., the perception of its own status and the perception of external stimuli, respectively). We will learn about a few technologies for building sensors to the level that suffices to understand their working principle and the information that they provide when measuring a physical quantity. The strategies for putting all the ingredients together are also part of this chapter, with the typical architectures for controlling the overall behavior of a complete robot with sensors and actuators. Going through such well-known robotics technologies and techniques is meant to help engage our understanding of the new challenges that soft robotics is bringing.

2.2   Robot Mechanics

Robotics is about movement, and so we start our dive into robotics basics with the fundamental elements that make a robot move: its mechanisms and actuators. Let us pose our first brick in figure 2.1. These components already enable a robot to interact with its environment.

A classical robot is typically composed of a chain of rigid links connected to each other by joints that allow their relative motion while keeping them in contact. In a classical configuration, like an industrial manipulator, the robot is an open kinematic chain, with one end fixed to a world reference and the other one, the end effector, free to move in space. Figure 2.2 provides a visual anchor to this general idea.

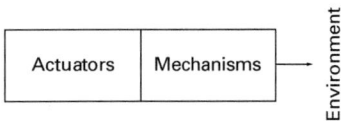

Figure 2.1
Robot mechanisms and actuators represent the first elements of a robot.

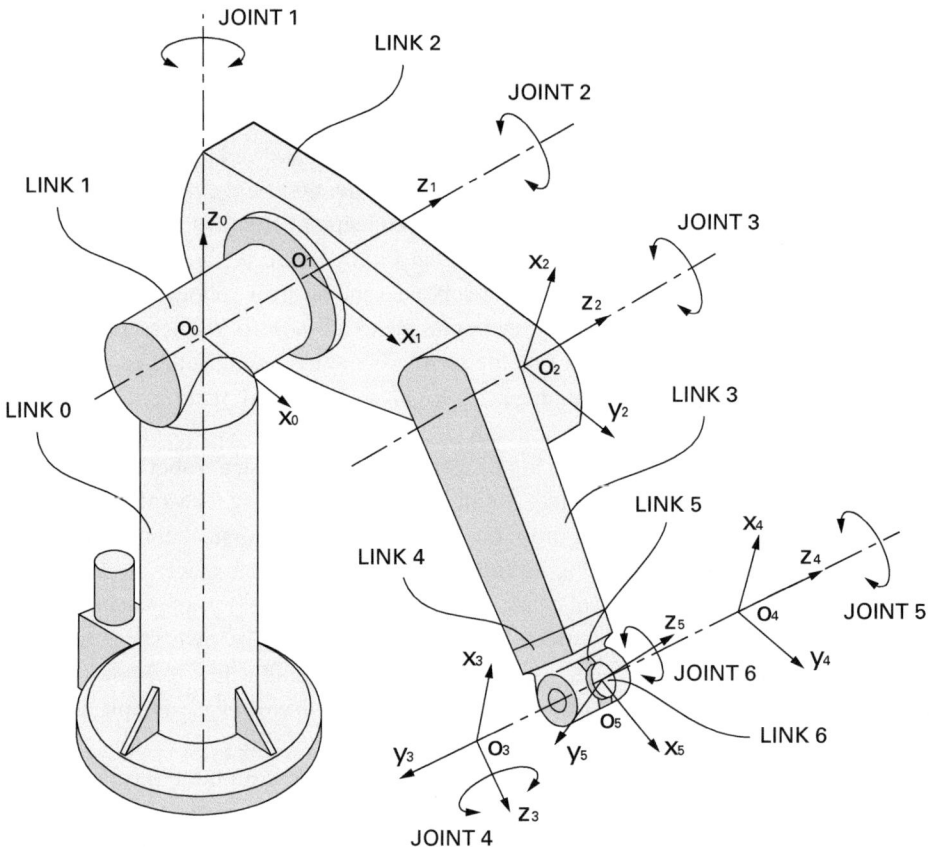

Figure 2.2
Conceptual view of a robotic arm, with a chain of rigid links connected with revolute joints.

Figure 2.2 probably gives you a familiar image of a robot, where the possible movements are rotations of a link with respect to the previous one. In this case, between two links we have a revolute joint, whose position is fully described by an angle θ. This is not the only possible way to connect two links. Translations of links are made possible by prismatic joints. In this second case, the joint position is fully described by a distance d. Depending on the combination of joint types (i.e., revolute or prismatic), we have diverse manipulator types. A robot like the one depicted in figure 2.2, with only revolute joints and no prismatic joints, is said to be anthropomorphic. You can easily tell why. Our arm joint movements may be approximated by rotations. On the other side, a robot with only prismatic joints is said to be Cartesian. That is easy to understand as well. Its movements can follow the directions of a 3D Cartesian reference frame. Other combinations of revolute and prismatic joints give other robot types. Together with the link geometries and ranges of motion, they determine the robot workspace, which is the space of the points that the robot can reach with its end effector. Some combinations of revolute and prismatic joints generate cylindrical or spherical workspaces, after which the robots are also named. A special case is the SCARA configuration, where two parallel, vertical revolute joints constrain the workspace on a plane and a final vertical prismatic joint set the plane height.

The position of a robot is expressed in two ways, or better in two spaces,[2] as explained below.

Robot position in joint space

Since each joint position is fully described by one value, either an angle or a distance, the set of all joint positions fully define the robot position, which can then be described by a vector of size n, where n is the number of joints:

$$\mathbf{q} = \begin{bmatrix} q_0 \\ \vdots \\ q_{n-1} \end{bmatrix} \tag{2.1}$$

Robot position in Cartesian space

Since the end effector is represented by a point, its position is used for expressing the robot position. Be reminded that the position of a point in the 3D Cartesian space is specified by both a 3D position and a 3D orientation:

$$\mathbf{x} = \begin{bmatrix} \mathbf{p} \\ \mathbf{\Phi} \end{bmatrix} = \begin{bmatrix} x \\ y \\ z \\ \phi \\ \theta \\ \psi \end{bmatrix} \tag{2.2}$$

where $\mathbf{p}$ is the vector of the end effector 3D position (x, y, z) and $\mathbf{\Phi}$ is the vector of its 3D orientation, with ϕ being the roll angle, θ the pitch angle, and ψ the yaw angle.

It is time to introduce another fundamental concept to keep in mind when taking the first steps in robotics: the concept of degree of freedom (DOF). It intuitively conveys the concept of movement extent. More precisely, it refers to the number of independent ways in which a body can move. In a robot such as the one in figure 2.2, the number of link-joint pairs determines the number of DOFs. Before reading further, try to think of how many DOFs you would need for your robot's end effector to reach all positions in its workspace. We will explore this later on.

Joints are where motion is generated, in a robot, and actuators are what generate such motion. Common actuators in robotics are electromechanical motors, like DC motors, but they are not the sole possible solutions. Reviewing the variety of existing actuators is beyond the scope of this chapter (instead, in chapter 4, we discuss the actuation technologies relevant to soft robotics). Here, let us assume we have an actuator on each joint. In case of revolute joints, we may even assume that the rotation axes of the joint and the motor coincide.

At this point, we know that each joint can move, and the robot consequently assumes a new position $\mathbf{q}$. The end effector will also reach a corresponding new position $\mathbf{x}$. What is the relation between $\mathbf{q}$ and $\mathbf{x}$? Very likely, the robot is commanded to reach a desired position $\mathbf{x}$ with its end effector, in the Cartesian space, regardless of the joint configuration. Since the motors are instead on the joints, how can we derive the joint positions to appropriately activate the motors? The answers to those questions are the kinematic transformations, in forward and inverse directions.

Forward (or direct) kinematics

Given a position $\mathbf{q}$ in joint space, the corresponding position in Cartesian space $\mathbf{x}$ is uniquely determined. The function to obtain $\mathbf{x}$ from $\mathbf{q}$ is the forward kinematic transformation K:

$$\mathbf{x} = K(\mathbf{q}) \tag{2.3}$$

Inverse kinematics

Given a desired position $\mathbf{x}$ in Cartesian space, the corresponding position in joint space is given by the inverse kinematic transformation K^{-1}:

$$\mathbf{q} = K^{-1}(\mathbf{x}) \tag{2.4}$$

Both transformations make use of a few geometric link parameters and transformation matrixes that are built up from those parameters. If you look back at figure 2.2, you see how reference frames can be set on each joint. Taking two consecutive joints, we can express the position and orientation of a reference frame with respect to the previous one. It is given by the distance of the frame origin from the previous frame origin and by the rotations of its three axes with respect to the corresponding axes of the previous frame. The transformation between the two can be described by a homogeneous transformation matrix $\mathbf{A}$ that contains the necessary rotations and translations:

$$\mathbf{A} = \begin{bmatrix} \mathbf{R}_{3\times3} & \mathbf{T}_{3\times1} \\ 0\ 0\ 0 & 1 \end{bmatrix} \tag{2.5}$$

Rotation matrixes depend on the axis and angle of rotation, as shown here:

$$\mathbf{R}_{x,\alpha} = \begin{bmatrix} 1 & 0 & 0 \\ 0 & \cos\alpha & -\sin\alpha \\ 0 & \sin\alpha & \cos\alpha \end{bmatrix}, \ \mathbf{R}_{y,\beta} = \begin{bmatrix} \cos\beta & 0 & \sin\beta \\ 0 & 1 & 0 \\ -\sin\beta & 0 & \cos\beta \end{bmatrix},$$

$$\mathbf{R}_{z,\gamma} = \begin{bmatrix} \cos\gamma & -\sin\gamma & 0 \\ \sin\gamma & \cos\gamma & 0 \\ 0 & 0 & 1 \end{bmatrix} \tag{2.6}$$

The translation component has the form of a 3×1 vector: $\mathbf{T} = \begin{bmatrix} d_x \\ d_y \\ d_z \end{bmatrix}$

By repeating the transformations between all joint pairs up to the end effector, we generate n transformation matrixes. By multiplying them, we obtain one final transformation matrix to express the end effector position in the base reference frame. A popular method to build the transformation matrix is the one proposed by Denavit and Hartenberg, which is based on two joint geometrical parameters and two link geometrical parameters and consists of the following steps:

1. For each joint i ($i = 1$ *to* n), set a reference X_{i-1} with

   • z_{i-1} axis corresponding to the i joint axis

   • x_{i-1} axis normal to the z_i and the z_{i-1} axes (free choice for $i = 1$) and frame origin O_{i-1} at the intersection with z_{i-1} (also free choice for $i = 1$)

   • y_{i-1} axis completing the frame according to the right-hand rule.

2. Set a reference frame on the end effector, with outgoing z_n axis.

3. For each joint i and for each link i connecting joints i and $i + 1$ ($i - 1$ *to* n), identify the four geometrical parameters as follows:

   • a_i is the minimum distance of the two z_{i-1} and z_i joint axes, measured along the normal line normal to both.

   • α_i is the angle between the two z_{i-1} and z_i joint axes, on a plane normal to a_i.

   • d_i is the distance between the frame origin O_{i-1} and the normal line above, along the z_{i-1} axis.

   • θ_i is the angle between the x_{i-1} and x_i axes, on a plane normal to the z_i axis.

4. Build a transformation matrix $^{i-1}\mathbf{A}_i$ between two consecutive joints $i - 1$ and i, through the following rotations and translations that bring X_i to overlap with X_{i-1}:

   • Rotate around x_{i-1} of an angle α_i, so as to align the z axes

   • Translate along x_i of a distance a_i, so as to align the frame origins

- Translate along z of a distance d, so as to overlap the origins
- Rotate around z_{i-1} of an angle θ_i, so as to align the x axes

5. Multiply all the transformation matrixes to obtain the transformation ${}^0\mathbf{A}_n$ from the end effector to the base.

Please note that this method only uses four parameters that, being geometric, do not change. One of the four is the joint variable (i.e., the joint angle θ in a revolute joint or the joint distance d in a prismatic joint).

Now we have the tools for transforming a position in Cartesian space into a joint configuration. The position in Cartesian space is a vector of always the same size: six elements, three for position and three for orientation. In joint space, instead, the size of the vector expressing the robot position varies with the number of joints n. For this reason, six is a magic number in robotics. When the number of joints is larger than six, the robot is redundant and can reach positions in the workspace with more than one joint configuration. With fewer than six joints, reaching all workspace positions is not guaranteed. So far, we have assumed each joint has an actuator producing its motion. In some cases, an actuator may move more than one joint—for instance, when motion is transmitted mechanically by cables. In those cases, robots are said to be underactuated, a concept that is relevant to soft robotics, as discussed later on.

Similar transformations are possible in the velocity spaces of the joints and the end effector, and they assume the name of differential kinematics. In this case, the transformation matrix is no longer composed of geometric, constant parameters but depends on the current robot configuration $\mathbf{q}$. It is known as Jacobian matrix $\mathbf{J}(\mathbf{q})$. It enables forward and inverse differential kinematics, as expressed in equations 2.7 and 2.8, respectively, where the dot notation indicates a derivative in time:

$$\dot{\mathbf{x}} = \mathbf{J}(\mathbf{q})\dot{\mathbf{q}} \tag{2.7}$$

$$\dot{\mathbf{q}} = \mathbf{J}^{-1}(\mathbf{q})\dot{\mathbf{x}} \tag{2.8}$$

The torques τ ultimately provided to the actuators are given by the following dynamic model of the robot:

$$\tau = \mathbf{B}(\mathbf{q})\ddot{\mathbf{q}} + \mathbf{C}(\mathbf{q},\dot{\mathbf{q}})\dot{\mathbf{q}} + \mathbf{F}\dot{\mathbf{q}} + \mathbf{g}(\mathbf{q}) \tag{2.9}$$

where $\mathbf{B}$ is the inertia matrix, depending on current configuration $\mathbf{q}$ and going with the joint accelerations; $\mathbf{C}$ is a matrix accounting for Coriolis forces that depend both on the current configuration and on the joint velocities $\dot{\mathbf{q}}$ and is multiplied by the joint velocities; $\mathbf{F}$ is the matrix of frictions and multiplies the velocities; and $\mathbf{g}$ accounts for gravity, depending on current configuration.

2.3   Robot Proprioceptive Sensors

Before moving on to robot control, we need to introduce the sensors that are instrumental to implement most control systems—namely, closed-loop controllers. Such controllers are based on a comparison between the target position and the current position. Therefore, robots need to be equipped with sensors that can detect their current position. Such position sensors are part of the robot proprioceptive sensors, using a term from biology (i.e., sensors of internal state). We learned that the joint configuration fully specifies a robot position. We are then going to see a few technological approaches for detecting the position of a joint and add another small brick to our conceptual scheme of a robot (figure 2.3).

It is also helpful to recall here the properties of a sensor (you find similarities with the actuator properties in section 4.3):

· **Transfer function** or **characteristic function**. The relation between the quantity to measure (input to the sensor) and the output of the sensor.

· **Linearity**. A measure of the deviation of the transfer function from a line.

· **Hysteresis**. The maximum output difference for a same input, depending on the fact that the input values are increasing or decreasing.

· **Accuracy**. The maximum error between the actual value and the sensor output.

· **Repeatability** or **precision**. A measure of the variability of the sensor output for a same input value.

· **Resolution**. The minimum variation of the input that generates a variation of the sensor output.

· **Sensitiveness**. The ratio between the output variation and the input variation.

· **Range of sensitivity**. The range of inputs detected by the sensor.

2.3.1 Mechanical Switches

Mechanical switches are very simple devices, where an external contact closes an electrical circuit, mechanically. The contact is then detected electrically and gives very simple, binary information: contact versus no contact. How can such limited information convey a position? In fact, such a device can detect one position only. In the case of a joint, a mechanical switch can tell when a joint has reached that specific position where the switch is placed. Despite their simplicity, mechanical switches are helpfully used as limit switch sensors to detect when a joint has reached the end of its range of motion.

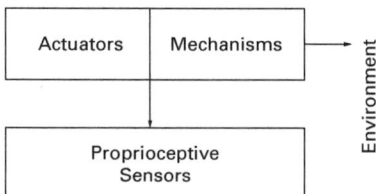

Figure 2.3
Proprioceptive sensors detect parameters of the robot current status. We are especially interested in position sensors, detecting each joint position.

2.3.2 Optical Encoders

Optical encoders are widely used in robotics, commonly integrated with the motors mounted on joints. As their name suggests, they are based on an optical device that consists of a light transmitter and a light detector. A disk with transparent and opaque slots is attached to the rotating axis to measure. The disk is positioned inside the optical device so that the opaque slots can interrupt the detection of light (see figure 2.4). When the disk rotates, the number of interruptions gives a measure of the number of slots passing by, which can be easily converted into an angle:

$$\theta = n\frac{360°}{R} \tag{2.10}$$

where n is the number of counted slots and R is the sensor resolution (i.e., the number of slots all around the disk).

2.3.3 Potentiometers

The electrical resistance of a resistor is proportional to its length. This simple principle allows one to measure a length by measuring a voltage change in an electrical circuit containing a resistor. According to Ohm's law, the voltage V is given by resistance R by current i:

$$V = Ri \tag{2.11}$$

Potentiometers are variable-length resistors, where the length is set by a moving wiper. The voltage measured in the circuit varies with the wiper movement, giving a measure of its position. While the voltage at the ends of the full-length resistor is given by equation (2.11), the voltage at the slider position is $V' = R'i$, with V' and R' being the measured

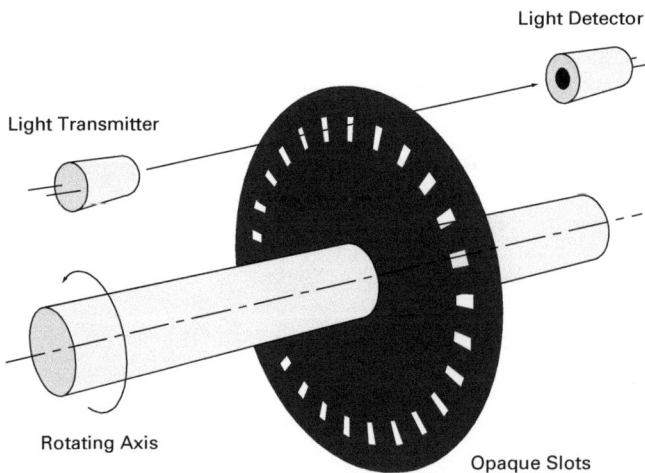

Figure 2.4
Working principle of an optical encoder.

voltage and the resistance at the wiper, respectively. With L being the full resistor length and L' the length at the slider position, $\dfrac{L'}{R'} = \dfrac{L}{R}$. From here, $R' = \dfrac{RL'}{L}$. The measured voltage V' is then represented as follows:

$$V' = R'i = \frac{RL'}{L}i = V\frac{L'}{L} \tag{2.12}$$

From there:

$$L' = \frac{V'L}{V} \tag{2.13}$$

The length L', which corresponds to the position we wish to measure, is found from the measured voltage V', the known initial voltage V, and initial resistor length L. The slide movement can be linear or rotational, providing potentiometers that measure lengths or angles (figure 2.5). In robotics, potentiometers are used to measure the position of a joint, either prismatic or revolute, with or without a motor mounted on it.

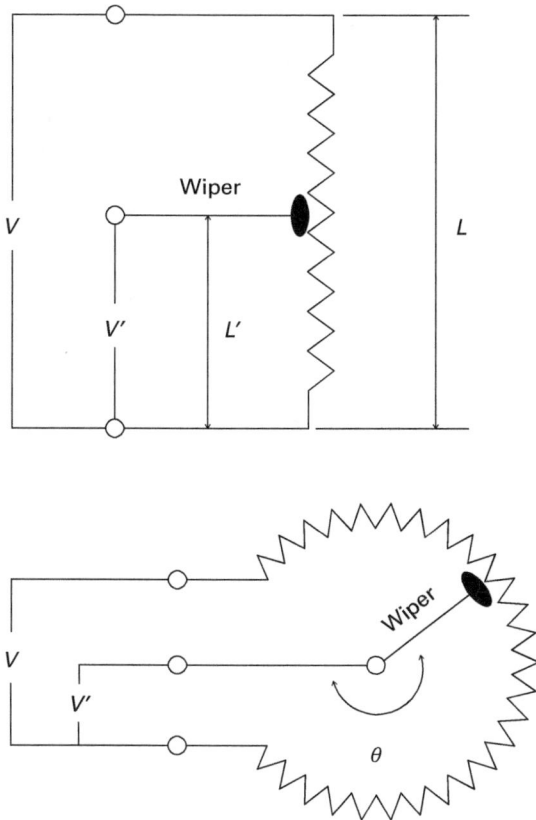

Figure 2.5
Working principle of (a) a linear potentiometer measuring the length L' and (b) a rotational potentiometer measuring the angle θ.

2.3.4   Hall-Effect Sensors

In 1879, E. H. Hall discovered that a small transverse voltage is generated across a current-carrying conductor in the presence of a static magnetic field, given the accumulation of charges on one side. Named after him, the Hall effect is stated as follows: In a conductor where a current i flows, immersed in a magnetic field of intensity B, a voltage V originates in the direction that is normal to both the current and the magnetic field. Figure 2.6 shows the elements at play. Such voltage V is proportional to the intensity of the current i and the intensity of the magnetic field B, as well as a material Hall coefficient H, while it is inversely proportional to the material thickness t: $V = \dfrac{iBH}{t}$

According to the Hall effect, the voltage V will change if either the current i or the magnetic field B changes. Assuming we keep the current constant, by measuring V, we can detect a change in the magnetic field that our conductor is immersed in. Hall-effect sensors are miniature devices where the Hall effect is elicited by a magnet producing the magnetic field and an electric circuit introducing the current and the voltage. As soon as the magnetic field is perturbed, the voltage changes and the variation can be measured. A possible cause is the proximity of a ferromagnetic object. Hall-effect sensors are in fact proximity sensors, detecting an object in their vicinity, with no need for contact. To convert them into position sensors, small magnets, perturbing the Hall-effect sensor magnetic field, are placed on the part to measure its position. In the case of our robot joint, a Hall-effect sensor can be fixed on one of the links, and small magnets can be placed on the other, moving link. The Hall-effect sensor detects the magnets passing by and can find the desired

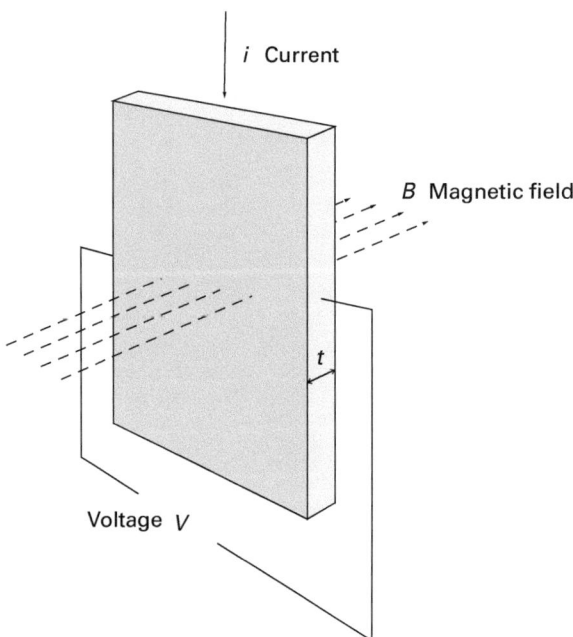

Figure 2.6
Hall effect.

position by counting them, with a strategy that is similar to the one used with optical encoders.

2.4 Robot Control

We have mechanisms, actuators, and position sensors. We are ready to build a control loop and make our robot move. We are building a control system where the desired position is pursued by nullifying the error with the current one, forming a closed loop, as depicted in figure 2.7. Generally, the input to the controller is a desired robot position, expressed in Cartesian space (i.e., a vector x_d). The output is a control signal $\mathbf{u}$ for all actuators, still in the form of a vector of size n, where n is the number of actuators that we assume here is the same as the number of joints. You may already see that some transformation will be needed, since the input is in the Cartesian space and the output is in the space of actuators or joints. Before seeing the two main control strategies, in joint space and in task space, we see how the motion of a single joint can be controlled, once we have a desired position for it.

2.4.1 Single Joint Control

When controlling the motion of a single revolute joint, the desired position is a joint angle θ_d. We assume a revolute joint with a motor on it that integrates an optical encoder as a position sensor. Generally, we also have a reduction mechanism between the motor and the joint, reducing motor speed for increased torque at the joint. Figure 2.8 shows this typical configuration.

We are interested in what is inside the controller box of figure 2.8, which is detailed in figure 2.9. You see two inputs: the desired joint angle θ_d and the current joint angle θ. The first component of the controller is a comparison of the two using a simple difference that represents the error. This error e is used to compute a control signal u, which becomes an input to the motor by power amplification. A typical strategy is the PID control, which determines u by using three components: a component that is *proportional* to the error (P), a component that accounts for the *integral* of the error (I), and a component that considers the error trend through its *derivative* (D). u is then calculated by the sum of these three components:

$$u = K_\mathrm{p} e + K_\mathrm{d} \frac{de}{dt} + K_\mathrm{i} \int e \, dt \tag{2.14}$$

An analysis of PID control, as well as other possible approaches, is beyond the scope of this chapter and this book. Intuitively, you may agree that we wish to correct the error

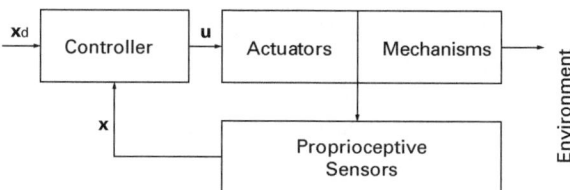

Figure 2.7
Elements of closed-loop robot control.

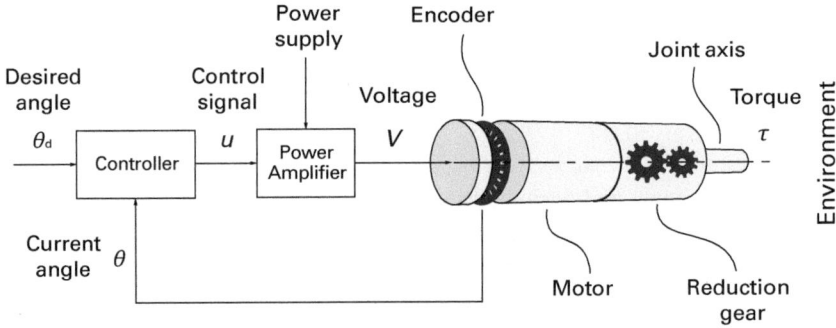

Figure 2.8
Typical configuration for the control of a single joint. The desired angle θ_d is the input to the controller, together
with the current angle θ. The controller computes the command for the motor, as a control signal u, which is
amplified to be a voltage V. The motor produces the torque τ for the joint axis rotation, with proper mechanical
reduction. The motor includes an encoder to give the current position to the controller as a feedback.

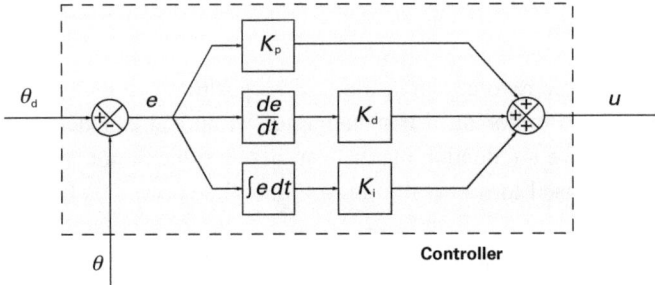

Figure 2.9
PID controller. The error is multiplied by a proportional parameter and summed to its derivative multiplied by
a derivative parameter and to its integral multiplied by an integrative parameter.

proportionally to how big the error is. This way, we risk overshooting, and the error may
change sign. Continuing to correct proportionally to the error will then generate oscillations. Consider, however, that the error derivative mitigates oscillations, because corrections also take into account the error trend, not only its magnitude. If the error is decreasing
already, the correction needed is smaller. The integral part ensures the error is nullified.
The PID controller is widely used, with different methods for setting the K parameters,
both theoretically and empirically.

2.4.2   Joint Space Control

Once we know how to make a joint assume a desired angular position, why not transform
the desired $\mathbf{x}_d$ position into the corresponding $\mathbf{q}_d$ position in joint space? The kinematic transformations discussed in section 2.2 can help. An inverse kinematic transformation brings the
target position from Cartesian space to joint space, right at the start of the control system and,
in fact, outside of the control loop (see figure 2.10). The error that the control system has to
process is in joint space, $\mathbf{q}_e$. It can be found by using the current position $\mathbf{q}$ as given by the
position sensors on the robot.

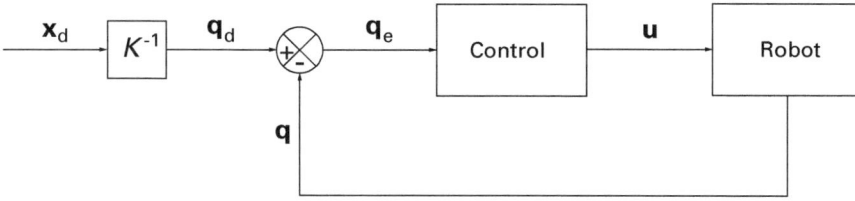

Figure 2.10
General control scheme in joint space.

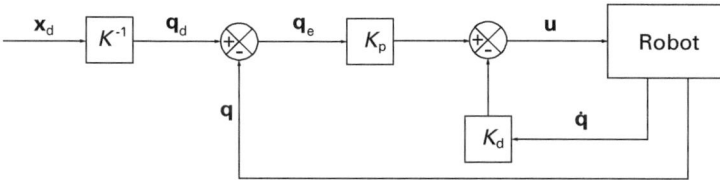

Figure 2.11
Controller in joint space with proportional component and a velocity control loop.

The control box of figure 2.10 may contain a diversity of computations. A simple example is computing **u** as proportional to $\mathbf{q}_e$, with a suitable $\mathbf{K}_p$ parameter:[3]

$$\mathbf{u} = \mathbf{K}_p \mathbf{q}_e \qquad (2.15)$$

As seen for the single joint control, a proportional component brings overshooting and oscillations. Taking joint velocities into account, with a proper $\mathbf{K}_d$ parameter, can mitigate oscillations by reducing corrections for high speeds:

$$\mathbf{u} = \mathbf{K}_p \mathbf{q}_e + \mathbf{K}_d \dot{\mathbf{q}} \qquad (2.16)$$

With these two components only, the controller becomes the one illustrated in figure 2.11.

Again, surveying all possible joint space controllers is beyond our scope, but the general concept is relevant to soft robot control in chapter 6. Computationally, the strategy of controlling the robot in joint space is convenient, since there is one inverse kinematic transformation outside of the control loop. However, the loop is closed on the joint positions, meaning that only the joint trajectories are controlled, irrespective of the end effector path. Reaching the final position is guaranteed, since it is exactly the same as the final joint configuration, but the positions assumed by the end effector to reach it are not controllable. Positions include orientations, which can be critical—what if a robot is handling a glass of water?

2.4.3   Task Space Control

In most cases, we need to control the trajectory followed by the end effector. We need to close the control loop on the error $\mathbf{x}_e$ between the $\mathbf{x}_d$ desired position and the **x** current position, both expressed in Cartesian space. The controller is said to work in task space (i.e., the space of robot operation). The current position is still given by the same position sensors on

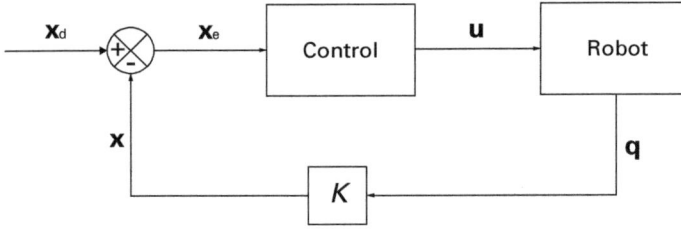

Figure 2.12
General control scheme in task space.

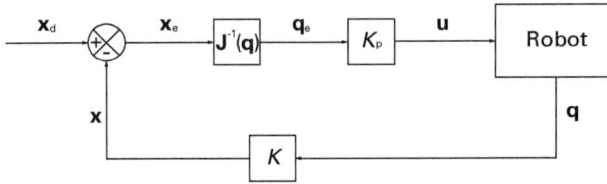

Figure 2.13
Control scheme in task space with an inverse Jacobian and a proportional component.

the robot joints, or actuators, but they still provide a **q** position in joint space. An additional step is necessary to transform it into an **x** position in Cartesian space. Kinematics transformations help again, with a forward transformation this time. See figure 2.12 for this general control loop. More importantly, inside the control box, we need to have more transformations, since the output **u** is in the space of actuators, or joints. Intuitively, they are inverse transformations and computationally expensive. Furthermore, they include joint and end effector velocities, so differential kinematics is involved, where the transformation matrixes depend on the position itself, **q**.

Another simple example may help us grasp the difference with control in joint space, especially from the computational viewpoint. In the control scheme shown in figure 2.13, the error in task space is transformed into an error in joint space by using the inverse Jacobian matrix. From equation 2.8, we can write the following:

$$\frac{d\mathbf{q}}{dt} = \mathbf{J}^{-1}(\mathbf{q})\frac{d\mathbf{x}}{dt} \tag{2.17}$$

For small displacements, we can derive the following:

$$d\mathbf{q} = \mathbf{J}^{-1}(\mathbf{q})d\mathbf{x} \tag{2.18}$$

Equation 2.18 holds for the error $\mathbf{x}_e$, which can be transformed into an error, $\mathbf{q}_e$, so as to finally have the follow equation:

$$\mathbf{u} = \mathbf{K}_p d\mathbf{q}_e = \mathbf{K}_p \mathbf{J}^{-1}(\mathbf{q})d\mathbf{x}_e \tag{2.19}$$

Figure 2.13 shows the control systems graphically and the computational steps inside the control loop, which include an inverse Jacobian and a forward kinematic transformation. Control in task space is generally more computationally intensive but allows for controlling the end effector trajectory.

2.5 Robot Exteroceptive Sensors

The next brick added in figure 2.14 connects the robot with the external world, not just for acting on it, which we saw was already possible with the first bricks of figure 2.1, but for receiving inputs and perceiving the surrounding environment. Though straightforward for human beings, this function brings a little revolution in a robot, since it makes it able to control its behavior according to what it perceives. Exteroceptive sensors complete the picture of a robot that we started from, defined as an autonomous system that not only can act on but also sense its environment. In robotics history, exteroceptive perception represents a second stage of development that was first characterized by robots used in factory settings and programmed with solid controllers that make them move, accurately and fast, within a structured environment that does not require exploration.

Exteroceptive sensors in a robot detect and measure physical quantities that describe the external environment. A robot can detect many physical quantities and is equipped with many sensors. Some of them are similar to human sensory systems, like vision and touch, or those of animals, like sonars. In the next subsections, a small selection of sensors is described by explaining their working principles. The selection is based on their relevance to the behavior of robots and especially soft robots.

2.5.1 Robot Vision

Vision is definitely an important sensory modality for a robot. Robots may be equipped with vision by using cameras and the well-developed techniques for image processing. Computer vision is a discipline per se and is not going to be covered in this book. Also, no major differences are worth mentioning in the use of vision in soft robotics.

2.5.2 Force/Torque Sensors for Robots

Force cannot be measured directly. It is measured indirectly—for example, by measuring strain.[4] A deformation is in fact a good indication of an applied force. Strain can be measured by taking advantage of the piezoresistive effect, by which conductive materials change their electrical resistance with strain. In a simple electric circuit, according to the Ohm's law $V = Ri$ (see also section 2.3.3), the change of resistance can be detected by a change in the voltage. This simple principle leads us to the strain gauge, a widely used

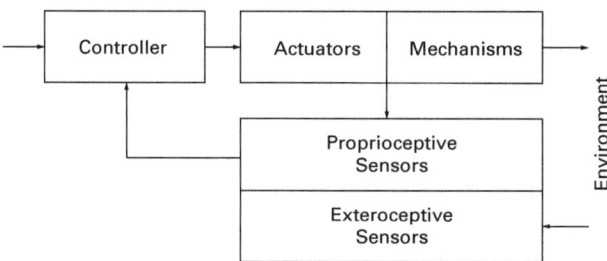

Figure 2.14
Exteroceptive sensors connect the robot to the external world in the reverse direction, allowing the robot to perceive external inputs.

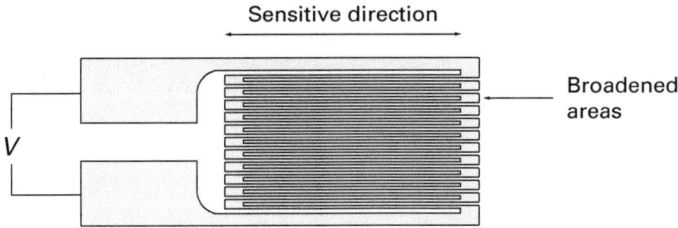

Figure 2.15
Strain gauge. The thin, long geometry of conductive elements favors their deformation along the sensitive direction. Broadened areas at the end reduce transverse deformations. Many other geometries exist that follow this principle.

miniature device that measures strain along one direction. The geometry of a strain gauge is designed in a way to amplify deformations in the sensitive direction while reducing them in the other directions (see figure 2.15).

A strain gauge is the basic element for building sensors that can detect forces and torques in three dimensions. When mounted on mechanical structures, a strain gauge filters one component of the applied force. The combination of opportunely arranged strain gauges can provide the three components and allow us to detect the force vector, as well as three components of torque. A typical configuration is the cross-shaped sensor with four arms departing from a central pillar, with strain gauges mounted on the arms. When a force is applied on the pillar, the vertical component acts on all arms equally, and the strain gauges respond in the same way, proportionally to the force intensity. Horizontal force components act on the arms differently, and the combination of strain gauge responses give their direction and intensity. See figure 2.16 for details.

Force and torque sensors are widely used in robotics. They are available in a variety of sizes and are typically mounted on the wrist of robot arms to measure the force and torque at the end effector.

2.5.3   Robot Tactile Perception

Tactile perception is a complex combination of sensory modalities and sensing elements, yet it is still far from the richness and density of the sense of touch of living beings. Artificial tactile sensors are arrays of sensing elements for detecting contacts, pressure, or forces. The most elementary information they can provide is whether there is a contact or not (see also the mechanical switches in section 2.3.1). This simple binary information can be arranged in an array and provides a tactile image with contact/noncontact areas, with no information on the intensity of the contact. More properly, tactile sensors provide information on the applied pressure or the force, which can be limited to its magnitude along the normal direction or may include the force direction in two dimensions (2D) or three dimensions (3D). Arranged in arrays, they provide tactile images.

The methods for detecting contacts and force span a range of technologies, from resistive to capacitive to optical and more. A few possible approaches and working principles are described in the following subsections. More technological approaches, specifically suited for soft robotics, are described in section 4.5.

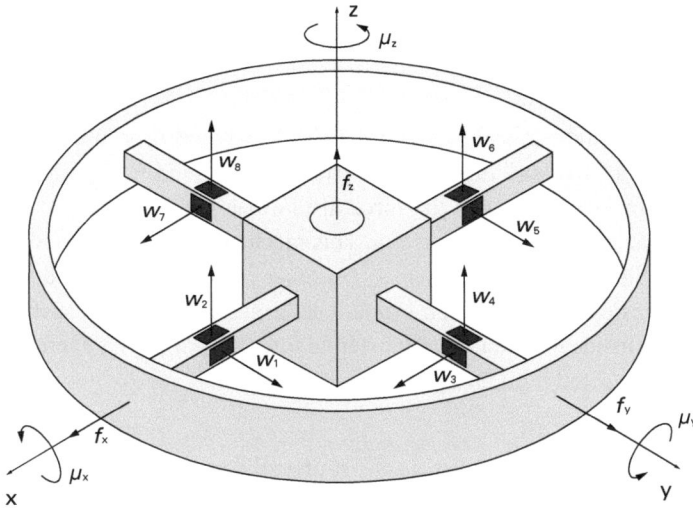

Figure 2.16
A cross-shaped force/torque sensor. A total of eight strain gauges are mounted on the four arms. The vertical force component f_z tends to bend the four arms equally, and the strain gauges w_2, w_4, w_6, and w_8 provide outputs while the others are silent since they are not deformed. The f_x component tends to bend two arms equally, where w_3 and w_7 respond while the f_y component tends to bend the other two arms, with response from w_1 and w_5. Similarly, a vertical torque acts on all arms equally, and it is detected through w_1, w_3, w_5, and w_7, while x and y torque components are detected by w_4 and w_8, and w_2 and w_6, respectively. Although strain gauges may be mounted on the mechanical structure in any position, the smart arrangement of this example separates the 3D components nicely, so that only a few readings are involved in processing the force and torque components. The translation from the strain gauge readings to six force and torque components is obtained through a coefficient matrix $\mathbf{C}$, typical of the sensor, such that $\begin{bmatrix} \mathbf{f} \\ \mathbf{\mu} \end{bmatrix} = \mathbf{C} \times \mathbf{w}$, where $\mathbf{f}$ and $\mathbf{\mu}$ are the vectors of three force and torque components, respectively, and $\mathbf{w}$ is the vector of the strain gauge readings. In this example, most $\mathbf{C}$ coefficients are zero.

Force sensing resistors (FSRs)

An FSR sensor is made of a conductive polymer that changes its resistance when a force is applied on its surface. A polymer film or ink is laid on a resistive film in rows and columns. An applied force makes the film resistance change by three orders of magnitude. This resistance fall can be measured through the voltage change in an electric circuit containing the FSR.

Capacitive tactile sensors

Two electrodes on the opposite surfaces of a polymeric film create a capacitor that, when the electrodes are pressed to get closer to each other, increases the electric field. The capacitance C depends on some geometrical parameters and material properties as follows:

$$C = \varepsilon \frac{A}{t} + C_f \tag{2.20}$$

where ε is a characteristic of the material; A and t are the area and thickness of the polymeric film, respectively; and C_f is a capacitance contribution from the edges of the electrodes, which is negligible when $A \gg t^2$. Since human fingers are conductive, they can play the role of one of the two electrodes in touch devices. Again, this change can be measured through a voltage.

Optical tactile sensors
An optical device, consisting of a light transmitter and a light detector, can be used in a tactile sensor that is designed in a such way that the light path is interrupted when a force is applied to it. This happens, for instance, when the force displaces a pin that reaches the optical device in between the transmitter and the detector. This mechanism can even modulate the amount of light, depending on the force intensity. Another possible approach does not require interrupting the light flow that is created inside a deformable medium in such a way that reflection is completely inside the medium. An external force deforming it generates external reflections that can be detected with an optical device.

Quantum tunneling composites (QTCs)
Using the quantum tunneling effect, these composites consist of a silicon matrix with nickel particles that never come into contact but get close enough to have electrons tunneling between them when the silicon matrix is deformed. This way, the QTCs transition from being virtually perfect insulators to metal-like conductors when a deformation, such as compression, twisting, or stretching, occurs. The transition follows a smooth and repeatable curve, with resistance falling from around $10^{13}\Omega$ to less than 1Ω. QTC sensors can have a variety of shapes and sizes.

2.5.4   Distance Sensors in Robotics

Another important sensory modality that is not going to be covered in this book is the perception of objects in the vicinity of robots, such as obstacles or environmental features that can be detected with ultrasound sensors, laser scanners, or other noncontact technologies. Although this sensory modality is especially relevant for mobile robots and navigation, it is not necessary for our soft robotics journey.

2.6   Robot Architectures

We now have a complete robot equipped with sensors, actuators, and a control system. We can take a step forward and control its overall behavior by using perception as an input and, to close the loop, taking action in response to the environment. As an example, a robot can use its cameras to detect an object and then compute from the images the point that the end effector needs to move to. When we refer to robot architectures, we mean the way the basic functions of sensing (SENSE), planning (PLAN), and acting (ACT) are connected in a robot. Before reading further, please take a moment and think of possible ways they can be connected (i.e., which function with which and in which direction).

2.6.1   Hierarchical Architectures

A straightforward way to connect SENSE, ACT, and PLAN is the SENSE-PLAN-ACT loop, closed on the environment. Sensors detect physical quantities from the environment (an image, for instance) and send this sensory information to a planning module, when then decides what action is necessary based on the sensory input. The planning module would

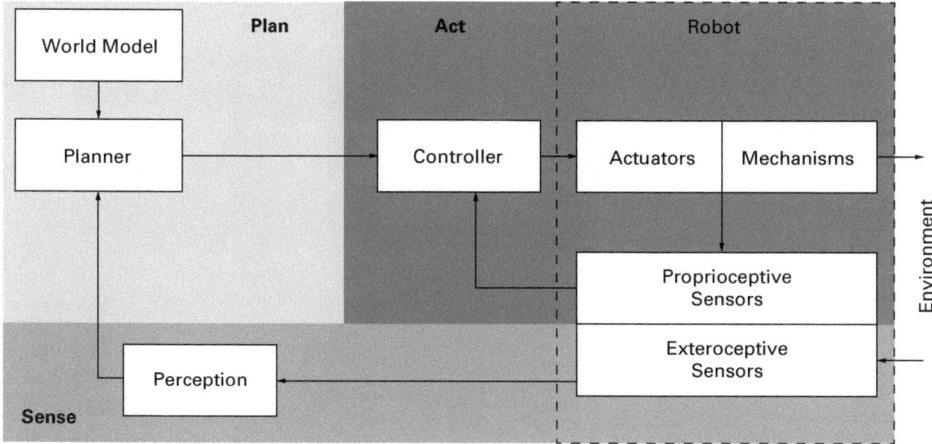

Figure 2.17
Hierarchical architectures. The information flows from the environment to the sensory system, which sends processed sensory inputs to the high-level controller or planner (top of hierarchy), which also uses a world model to generate the motor commands for the mechanical system, which acts back on the environment. At a lower level, a controller uses proprioceptive feedback for robot motion.

then send the commands to the motor system (i.e., robot actuators and mechanisms), which acts back on the environment. This is in fact the classical structure of *hierarchical* architectures, depicted in figure 2.17. Please note that the planner needs some a priori knowledge of the world, of the robot itself, and of some policies that a programmer may wish to set. All this is contained in a world model, which is part of hierarchical architectures. The direction of connections is clear from figure 2.17 as well.

2.6.2   Reactive Architectures

Inspired by the behavior of small animals, especially insects, in the 1980s, Rodney Brooks proposed a disruptively different approach by questioning the need for planning at all. By arguing that there is no need for a world model if the robot has sensors for detecting it, he demonstrated that a direct connection between sensors and actuators can be enough to generate a seemingly intelligent behavior, reactive to external stimuli. For this reason, architectures implementing this approach are known as reactive architectures (figure 2.18). The PLAN module is fully removed from the picture, together with the world model connected to it.

In fact, in a different context years before, with no computers or robots, Valentino Breitenberg demonstrated that idea. As an experimental psychologist, he wanted to show that a behavior may look like intelligence, guided by reasoning and decision-making, while it can even be produced by direct connections between senses and a motor system, without any computation involved. He proposed simple vehicles with wheels actuated by motors and sensors responding to temperature. The first vehicle has one wheel only, and the motor speed is proportional to the sensor response; see figure 2.19(a). It means that if the sensor detects a higher temperature, the motor and the wheel run faster. The sensor and the motor are connected directly with an electrical wire, with no computing units. This vehicle cannot do much more than move straight and faster in warmer areas. With two wheels, the vehicle can steer

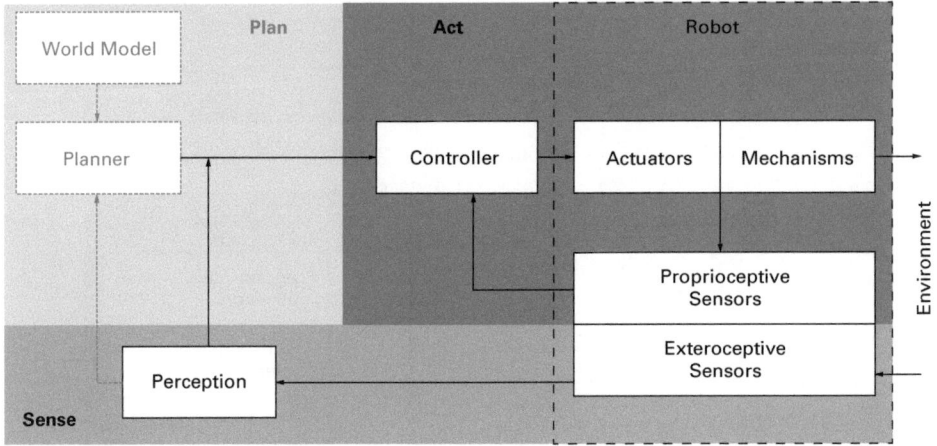

Figure 2.18
Reactive architectures. Perception is connected to action directly, without use of a world model or a planner.

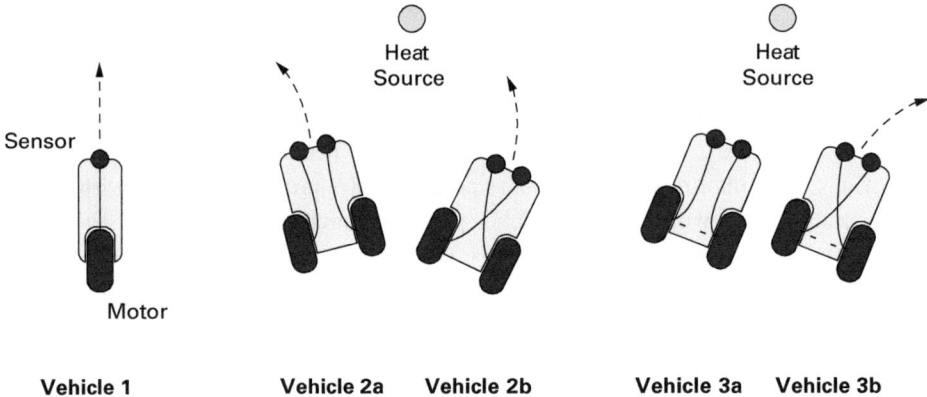

Figure 2.19
Braitenberg vehicles. (a) A vehicle with just one wheel and one sensor. (b) Vehicles with two wheels and two sensors, directly and inversely connected. (c) Vehicles with negative connections and the resulting opposite behavior.

when wheel speeds are different. And two sensors can make this happen. In the second vehicle, two wheels are connected to two sensors in the same way as in the first vehicle; see figure 2.19(b). At this point, if a sensor is closer to a heat source and detects a higher temperature, the wheel on its side runs faster, and the vehicle turns away from the heat source. This behavior was called *fear*. By crossing the connections—left wheel with right sensor and right wheel with left sensor—the behavior is opposite, since the opposite wheel runs faster and the vehicle turns toward the heat source. This behavior was called *aggression*. More complex Braitenberg's vehicles can be built by crossing or inverting the sign of connections—shown in figure 2.19(c)—and more behaviors can be generated that look as if they are guided by some form of reasoning when they are not.

In robotics times, behavior can be generated in a similar way by connecting the sensing system directly to the action system. This SENSE-ACT pair is said to be a *behavior*, and the use of reactive architectures is also known as *behavior-based robotics*. A SENSE-ACT pair is very simple code of a few lines that is easy to execute, so the robot is reactive to external stimuli in a shorter time than with hierarchical architectures involving a planner. More complex behaviors are built with a number of behaviors that run in parallel. This parallel execution of functions gives further advantages and reactivity. It makes the system modular, since behaviors are independent of each other and can be added or removed to add or remove functions, with no need to reprogram the others. A major drawback is that all behaviors compete to send commands to the same actuators. Suitable arbitration or combination methods are used to derive the actual motor commands. Rodney Brooks's subsumption architectures solve the concurrency of behaviors by introducing connections between them. A behavior can suppress the input to another behavior or inhibit its output. Independence is not lost, since all behaviors keep running in parallel, but some behaviors become ineffective thanks to this connection mechanism.

2.6.3 Hybrid Architectures

Since hierarchical architectures have their own merits in terms of behavior predictability and planning, hybrid architectures are often implemented in robotics, where a planner together with a world model trigger reactive behaviors. This way, the advantages of both approaches are capitalized on.

2.7 Summary

We have reached the end of this shallow yet intense dive into robotics. You learned, or recalled, the major concepts that are relevant for learning the soft robotics topics of the next chapters. We built our ideal robot by starting with its mechanisms and actuators and their basic relationships in kinematics and dynamics. We added proprioceptive position sensors before building closed-loop control systems that use their feedback. We saw the differences of controlling a robot in joint space or in task space, with corresponding required transformations. By adding exteroceptive sensors, we made our robot interact with its surrounding environments in two directions, and we saw how perception can be connected to action, with or without a planner.

Self-Assessment Questions

1. Why is six a magic number in robotics?

2. Apply the Denavit-Hartenberg method to a two-link robot with your own geometry.

3. What is the minimum angle measured by an encoder with resolution $R = 720$? Find a joint angle θ from the encoder reading $n = 1800$, taking into account a reduction ratio $k = 10:1$. If the same angle would be measured by a potentiometer of total length $L = 0.002$m, powered with $V = 5$V, what would the reading V' be?

4. What sensors would you use in a robot used for manipulation tasks?

5. What are the possible connections between SENSE, PLAN, and ACT modules?

Further Readings

Mordechai, Ben-Ari, and Francesco Mondada. 2018. *Elements of Robotics*. Cham, Switzerland: Springer.

Breitenberg, Valentino. 1986. *Vehicles: Experiments in Synthetic Psychology*. Cambridge, Mass: MIT Press.

Brooks, Rodney Allen. 1999. *Cambrian Intelligence: The Early History of the New AI*. Cambridge, Mass: MIT Press.

Corke, Peter, Witold Jachimczyk, and Remo Pillat. 2017. *Robotics, Vision and Control*, 2nd ed., Springer Tracts in Advanced Robotics (STAR, Vol. 147). Cham, Switzerland: Springer Nature.

Mihelj, Matjaž, Tadej Bajd, Aleš Ude, Jadran Lenarčič, Aleš Stanovnik, Marko Munih, Jure Rejc, and Sebastjan Šlajpah. 2019. *Robotics*, 2nd ed. Cham, Switzerland: Springer.

Murphy, Robin. 2000. *Introduction to AI Robotics*. Cambridge, Mass: MIT Press.

3 Bioinspiration and Biomimetics

Chapter Objectives

- To define bioinspiration and biomimetics
- To learn some specific principles to steal from nature
- To learn a methodology for designing bioinspired robots

3.1 Overview

This chapter is going to make you look at nature with different eyes. It will change your view of robotics, too. If that happens, the chapter will have fulfilled its purpose. You will learn what bioinspiration and biomimetics are, you will review a few well-known and some unexpected examples of them, and you will understand how they can help in robotics. In fact, we will see that the benefit is reciprocal, and biology can benefit from bioinspired robotics as well. We are going to discover a world of principles that can inspire your engineering and robotics work. Finally, we will formalize a method for designing bioinspired robots by making good use of the inspirational models that nature provides. This is important to correctly implement those bioinspired principles that form the foundation of soft robotics.

3.2 Why Bioinspiration and Biomimetics?

Life on earth is the result of a smart mechanism for adaptation. Living beings undergo the natural selection that Charles Darwin hypothesized and described in his writings in the nineteenth century. Small genetic variations in a population make some individuals better suited to the environment where they live so that they survive, reproduce more, and transmit their good genetic heritage to their successors. Natural selection is the key mechanism for evolution of a population, over generations. It helps the population fit to its own environment, especially when it changes. Here comes the amazing adaptation of species that constitute biodiversity on earth.

Humankind has long looked at nature to borrow ideas and smart working principles for human-made tools. If living beings are so well adapted to life on earth, why not steal their secrets when toolmaking? This is the underlying question in bioinspiration and biomimetics. In the next section, we are going to review a few examples of bioinspiration across many areas of human activity. Focusing back on robotics, the question then becomes: If we are aiming to design robots that work in natural and human environments, couldn't we also learn from the secrets of living beings who are so well adapted to their habitats? This question touches one of the fundamental issues that robotics cannot get rid of. The first robots were built for working in factories, which is what they still do excellently. So, they are designed for an environment that we can consider to be artificial, adapted to the robots themselves. We call it a *structured* environment. Despite the huge progress of service robotics—that is, the use of robots in many more environments beyond factories—the common image of a robot is still that of an industrial robot and its clumsiness in natural environments. We call them *unstructured* environments, by contrast. In fact, most of the technologies used in robotics, even in service robotics, are derived from industrial robotics techniques and based on the same conceptual approach. This is why roboticists as well have been looking at nature for solving some of the many problems that robots have to face when negotiating natural or human environments (i.e., unstructured environments).

3.3   What Are Bioinspiration and Biomimetics?

There are many words that are used to describe the concept that we are addressing in this chapter. It is good to know them and analyze their possible differences.

Biomimetics (bio = life, mimesis = imitation, in Greek) is a term with a wide scope. It includes the very general idea of imitating something from life. It then includes any living organism and any sector of human activity. *Biomimicry* is synonymous with biomimetics, with the same Greek origin, as well as the less common *biomimesis*.

Bioinspiration is a looser word referring to the different ways an engineered system is similar to a biological model. *Biologically inspired* is synonymous with it, so the adjectives bioinspired and biologically inspired are used with the same meaning.

Bionics is a word that merges biology and electronics. It was coined in 1958 by Jack E. Steele, a medical doctor in the US Air Force. It is sometimes also used to indicate the imitation of life in engineering. In any case, bionics is mainly used to indicate the integration of natural and artificial parts, as in the case of prostheses, which is outside our scope.

For most of the scope of this book, the terms bioinspiration and biomimetics are used as synonyms. However, you will learn a slight difference between the two terms at the end of this chapter, where we analyze our methodology for designing bioinspired and biomimetic robots.

Definition 5   Bioinspiration and biomimetics   Extraction of principles from the observation of living beings to adopt in building human-made products.

While it is important to agree on the terminology to use in this book, it is even more important to agree on a clear definition of those words. So, before we make our final word choice, let us go more in depth into the meanings.

At this point, you must be already convinced that taking lessons from nature is worthwhile. The argument we make is that living beings adapted to their habitats through evolution, thus those solutions may be are well suited to be adopted by artificial systems. How well suited, though? And for what? Living beings are the results of natural selection. Natural selection has two main goals: survival and reproduction. Individuals good enough to survive and reproduce will transmit their genetic patrimony to their successors. And this is what we observe as our models in bioinspiration and biomimetics. Please be aware that good enough is not optimal. Living beings are the product of incremental modifications for adaptation to a changing environment. They do not represent the optimal solution that engineers usually look for. And the biological goals of survival and reproduction are not exactly the tasks of our artificial systems. This is not an attempt to bring arguments against bioinspired approaches, of course. It is an attempt to focus on what we should beneficially look at, in nature. Blindly copying living beings is not a good approach to design bioinspired systems. Instead, we should be able to extract the principles that can help with our tasks at hand. It is not an easy task.

It is now time for a few examples in order to better understand what bioinspiration and biomimetics are and to test your ability to apply our definition to identify objects, and ultimately robots, that comply with it. You will find several objects from your daily life and some unexpected cases.

3.3.1   Hook-and-Loop

This bioinspired object is under our eyes daily. Attaching/detaching straps are an extremely useful solution for many of our daily tasks, adopted in clothes and shoes as well as in tech-nologically sophisticated professional equipment. It is a patented invention authored by a Swiss engineer in 1955, George de Mestral, who also originated the Velcro brand that com-mercializes it. When walking his dog on the grassy Swiss mountains, de Mestral observed that the dog used to bring back home a lot of plant burrs (from *Arctium lappa*, to be precise), attached to its fur. This precise strategy by which a plant would have its seeds spread over a large territory was quite annoying indeed, but fortunately the burrs could be easily detached by pulling them away from the dog's fur. Of course, the plant strategy ensures the seeds are released at some point. Interestingly, the burrs could even attach again. Using a microscope, de Mestral investigated the principle behind this behavior. He found the burr surface covered by microscale hooks, as shown in figure 3.1(a). From there, he had the idea of manufacturing both parts of this attachment system (i.e., a strip with micro-hooks and a strip with fibers resembling the dog hairs). He invented the hook-and-loop.

3.3.2   Lotus-Inspired Painting

If you have the opportunity to observe a lotus leaf floating on the water, you see that any drop falling on its surface keeps its rounded shape and rolls away without being absorbed by the leaf, as illustrated in figure 3.1(b). This hydrophobic behavior keeps the leaf surface very clean, despite the pond-like and potentially dirty habitat of the plant. When water drops roll away, they bring small dirt particles with them. The principle of interest to us here is the hydrophobic behavior called *lotus effect*, which depends on the microstructure of the lotus leaf surface. A hierarchy of microstructures formed by papillose epidermal cells makes the surface rough and reduces the contact surface available; epicuticular wax tubules provide a hydrophobic coating. Based on this principle, a a coating used to paint wallsis commercial-ized as Lotusan for self-cleaning building surfaces.

3.3.3   Shinkansen Bullet Train

When officially inaugurated on October 1, 1964, just 10 days before the opening of the Tokyo 1964 Olympic Games, the Shinkansen train (or bullet train, as it was nicknamed) was one of the fastest trains on earth, traveling at over 300 kilometers per hour (km/h). It was an extraordinary demonstration of technology, connecting remote parts of Japan with unmatched speed and convenience. Such unprecedented train speed made engineers discover a couple of problems that do not occur with traditional, slower, trains: tunnel boom and pantograph noise.

Tunnel boom
In exiting tunnels, the trains used to make a sudden and loud sound—a boom. That is given by the pressure difference inside and outside the tunnel. The engineer assigned to solve this issue, Eiji Nakatsu, observed that kingfisher birds, when hunting fish, enter the water without splashes. They are addressing a similar problem—namely, passing from a low-resistance medium (air) to a higher-resistance medium (water) without a boom, or a splash, that can reveal their presence to prey. Nakatsu investigated the birds' behavior and observed the principle that could be applied to the train: the streamlined shape of the bird's beak, the part that enters the water first. That is why today's Shinkansen trains have a streamlined front. See figure 3.1(c).

Pantograph noise
The Shinkansen pantograph, the part that connects the cars to the electrical lines above them, tended to be noisy as well because of the high speed of the air flow going through them. Again, Nakatsu looked at nature and observed owls, which can fly extremely silently and reach their prey, at very high speed, without getting noticed. He found a good principle that he could take to apply to his problem as well: the structure of the primary feather of owl wing, which generates small vortices. The pantograph was modified to a shape that creates small vortices instead of the large ones generated previously.

With his bioinspired solutions, Eiji Nakatsu made Shinkansen trains quieter and 10 percent faster, while consuming 15 percent less electrical power.

3.3.4   Eiffel Tower

You are going to be very familiar with this example, but you may not be aware that it is bioinspired. Under pressure to complete the design of the tallest building on earth for the Exposition Universelle of 1899 in Paris, engineer Alexandre Gustave Eiffel was struggling to find a suitable technical solution. The materials used at the time, like stones and iron, were too heavy, and a very tall building would have collapsed under its own weight. This problem is similar to one solved by nature millions of years ago. Animals need bones to stand against the force of gravity and to carry body weight, but bones should not add too much mass. This is especially true for birds, which need to be very light for flying even while their bones carry their body weight. The principle that Gustave Eiffel found in bones is that they are not a solid piece of tissue. Bones are composed of many connected beam-like *trabeculae*, leaving a large percentage of empty weightless space. See figure 3.1(d).

Indeed, trabeculae are not randomly oriented, but they tend to be oriented along the main lines of force acting on the bone so as to increase the bone's resistance to those forces. This is why the Eiffel Tower consists of iron pillars and beams oriented along the lines of force that gravity applies on it. And it is still standing, at 312 meters and 10,000 tons, over Paris.

3.3.5 Geckos and Robots

An amazing ability of geckos is that they can climb any surface, even very smooth ones, vertically or even upside down. They literally attach to the surface, but they do not use any form of natural gluing effect. Their adhesion is dry. The underlying principle relies on a hierarchy of structures on the gecko's toes, at different microscales, that generate attraction forces. Each toe has 10 lamellae, as you can easily spot with your naked eye in figure 3.1(e). From each of them, thousands of setae depart, with hundreds of spatulae on their ending surface. At this point, your naked eye does not really help. A spatula pad can be in the order of $100 - 200\,nm$, with a thickness of 5 to $10\,nm$. Their intimate contact with the contact surface generates high adhesion and high friction force. Another important principle is that the adhesion mechanism depends on the angle and direction of the feet as they approach the surface, which translates into an easy detachment if a simple feet movement reverts that angle of direction.

We need to travel to Mark Cutkosky's lab at Stanford again (after our trip in chapter 1), to see a gecko-inspired robot that can climb vertical surfaces. Its feet are made of soft materials that first of all make them conform to the contact surface. Then, they reproduce the lamellae structure on the sole of the feet. The soft toes are actuated so as to produce the detachment movement. The robot possessing such sophisticated feet, StickyBot, has been seen climbing a window glass of Stanford campus.

3.4 Brief History of Bioinspiration and Biomimetics

Bioinspiration and biomimetics are not new at all. Even if the words and the blooming of bioinspired research are relatively new, humans have been attempting to copy nature for a long time. We are going to see some of those attempts in the following subsections, jumping over many-year time spans.

3.4.1 Ante-Litteram Bioinspiration and Biomimetics

Silk is an incredible material produced from the cocoon of an insect, the silkworm. The unique properties of its fibers have made silk a valuable fabric since ancient times until today. It seems that the Chinese tried to produce artificial silk already 3,000 years ago. Although that early effort was not completely successful, it is considered to be the very first example of biomimetics in history. It is evidence of how much bioinspiration and biomimetics are deeply ingrained in human culture. And we had to wait until the nineteenth century to have the first fibers named as artificial silks (i.e., rayon and viscose).

Much later, in the fifteenth century, Italian artist and inventor Leonardo da Vinci looked at nature for his science and art. Initially, Leonardo approached the study of the human musculoskeletal system for artistic purposes. He was unsatisfied by the existing anatomical descriptions needed for painting human bodies. Little by little, he developed a passion for understanding muscular physiology and mechanics. He made detailed drawings and wrote descriptions where, to better explain the principles he discovered, he replaced the muscles

(a)

(b)

(c)

A

B

10 cm

(d)

Figure 3.1
(continued)

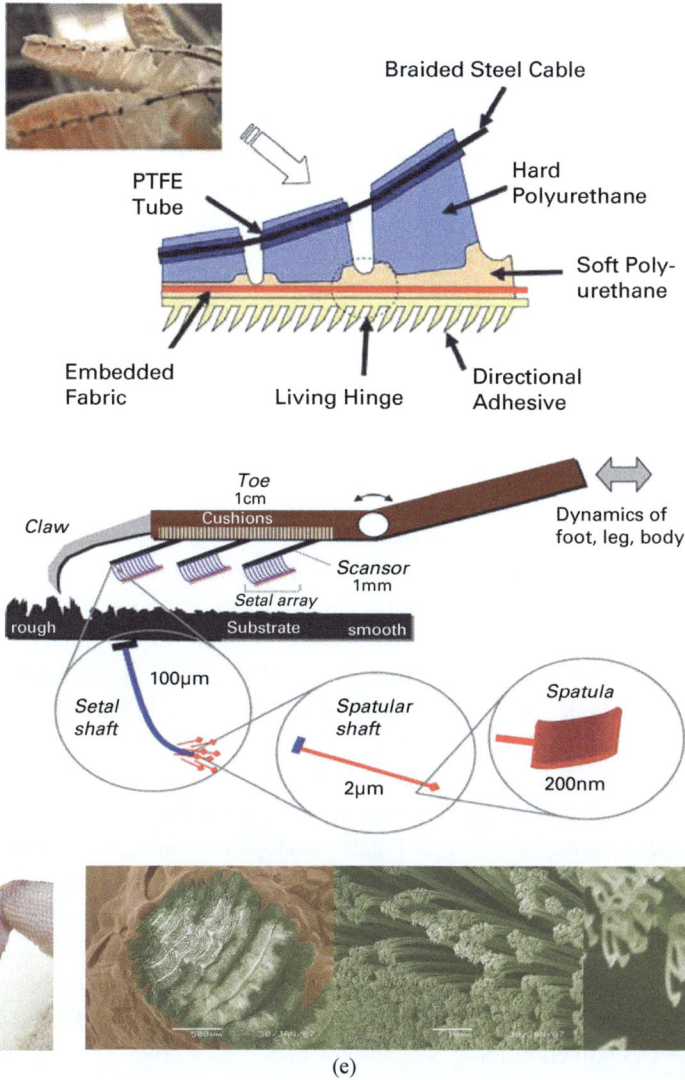

Figure 3.1
Examples of bioinspired objects and robots. (a) The hook-and-loop adhesion system (photo by Hadley Paul Garland via Flickr licensed under CC BY-SA 2.0) and the plant burr that inspired it (photo by Roger Culos via Wikimedia Commons licensed under CC BY-SA 3.0). (b) Lotusan painting (copyright Sto SE & Co. KGaA, reproduced with permission) and water drops on a hydrophobic lotus leaf (photo by GJ Bulte via Wikimedia Commons licensed under CC BY-SA 3.0). (c) The Shinkansen train front (photo by Mike Knell via Flickr licensed under CC BY-SA 2.0) and the kingfisher's beak that inspired it (public domain drawing from Sharpe 1868). (d) The Eiffel tower structure (photo by Ank Kumar via Wikimedia Commons licensed under CC BY-SA 4.0) and trabeculae in a human femur (Bishop et al. 2018, fig. 1). (e) A gecko-inspired robot foot (reproduced with permission from Cutkosky and Kim 2009) and a gecko foot (photo by FurryScaly via Wikimedia Commons licensed under CC BY-SA 2.0) and its hierarchical structure (photo by Oskar Gellerbrant via Wikimedia Commons licensed under CC BY-SA 2.0).

and tendons with ropes and wires. Notably, this approach resonates greatly with the second purpose of biorobotics explained in section 3.5—that is, explaining natural phenomena with technological analogues. He looked at nature for many of his inventions. He studied birds to understand flight and to design machines to make human beings fly. One may argue that he was not always able to build working prototypes of his machines, as in the case of the *Ornithopter* flying machine. Designed in 1485, it is a flapping-wing machine operated by a person through pedals. Was it a technological gap that prevented Leonardo from actually building it? Consider that his typical construction materials were wood and ropes. In 2010, the Human-Powered Ornithopter (HPO) Project team at the University of Toronto developed the flapping-wing machine with today's technologies. It sustained both altitude and airspeed for 19.3 seconds and covered a distance of 145 meters at an average speed of 25.6 km/h.

3.4.2   Birth and Growth of the Field

In 1957, a doctoral student in the United States attempted to reproduce the electrical action of a nerve in a physical device. He wrote a paper based on his work and its title included, for the first time, the word "biomimetics" (Schmitt 1969). The student, Otto Schmitt, is still considered to be the inventor of the word. The meaning goes beyond his specific study to include the transfer of ideas from biology to technology in general.

Later on, the term biomimetics entered Webster's Dictionary in 1974, defined as "the study of the formation, structure, or function of biologically produced substances and materials (as enzymes or silk) and biological mechanisms and processes (as protein synthesis or photosynthesis) especially for the purpose of synthesizing similar products by artificial mechanisms which mimic natural ones."

A milestone in the history of bioinspiration and biomimetics was a 1989 workshop held in a small village in Tuscany, Il Ciocco. It gathered together young and enthusiast scientists who wanted to advance bionics beyond its depiction in science fiction. Today, we can see that those first steps produced solid achievements and technological progress in both academic science and industrial technology. If we could attend the workshop with a time machine, we would see robotic elephant trunks, legged robots, biped humanoids, artificial hands, retina-like vision systems, human-inspired motor control, artificial muscles, and robot swarms. All evolved to today's examples of robotics progress.

Despite human beings representing a primary model of bioinspiration, especially in robotics, many animals also serve that purpose. Surprisingly, plants do as well. Plants have long contributed to biomimetics because of the variety of properties of their tissues, spanning from soft stems to hard and resistant branches and trunks, which inspired new materials and structures. More recently, plants have become models for robotics, too. You may think that there is a mismatch here since robotics is about movement and plants do not move, but this is a kind of misunderstanding. Plants move, in fact, but at a different timescale. Their movements are purposeful and very efficient, computationally and energetically. Both reasons make them a good model for robotics. A milestone in plant-inspired robotics is the PLANTOID project,[1] conceived and run by Barbara Mazzolai at the Italian Institute of Technology (IIT) with funding from the European Commission from 2012 to 2015. The main achievements are technologies for robots growing, by adding material, and for exploring soils in an efficient and high-resolution way.

An analysis of the growth of the contribution of bioinspiration and biomimetics to science and technology, in the broadest terms, is given in a *Nature* paper (Snell-Rood 2016) that highlights not only how successful the approach is, but its underused potential. We are just scratching the surface of bioinspiration and biomimetics, and we are studying a very limited number of species still. Also, the full potential can flourish if there is a higher involvement of biologists in biomimetics research. The method that you are going to learn in section 3.9 can respond to this purpose.

3.5 Biorobotics

There is a sort of beneficial side effect when you build an artificial system that implements principles taken from a biological model: you gain insight on the natural system itself. Implementing a principle observed in a living being can tell you whether the principle itself is correct. Some refinement loops are often needed to arrive at a fully correct principle and a correct implementation of it. And this may lead to new insights about the observed living beings and even to new discoveries.

A paradigmatic case is the study undertaken by Shigeo Hirose at the Tokyo Institute of Technology. As a young student in the 1970s, he was developing one of the first snakelike robots, so he studied the literature on snake locomotion undulatory movement. Initially, when reproducing this movement in his robot, it did not work. After struggling with his design and technology, he realized that an additional raising movement was necessary on the snake bends involved in locomotion. With that solution, at least his robot could crawl. He went back to the animals, then, and found that this is, in fact, what snakes do, which was not yet documented in the literature.

Along this line, in robotics, we consider a two-way relation between biology and technology. On one side, observing biological models provides principles for bioinspired robots that address real-world applications; on the other side, building biomimetic robots provides tools for studying biological systems. In 2001, Barbara Webb from the University of Edinburgh published a book that describes biorobotics and represents the contemporary step of our short history (Webb and Consi 2001).

The word *biorobotics* encompasses this double-face purpose; it even comprehensively includes the field of biomedical applications of robotics. In the words of Paolo Dario, a pioneer of modern robotics and biorobotics at the BioRobotics Institute of Scuola Superiore Sant'Anna in Pisa (Italy), biorobotics is both science and engineering because it is about inventing new technologies and discovering new knowledge. He gives a useful definition in a commentary published in *Science Robotics*: "science for robotics and robotics for science." This is now the motto of the journal after the editorial appeared in first issue authored by all the founding editors.

Definition 6 Biorobotics science and engineering Biorobotics keeps the living world (and thus life sciences) at its core, investigates different applications of bioinspired machines and robots, and validates scientific hypotheses. (Yang et al. 2016)

The biorobotics approach goes back to a time before robotics even existed. The technologies available at different times in the past were used to replicate principles observed in nature, not so much for the purpose of obtaining helpful machines but for investigating living beings.

In 1943, a milestone paper used the word "cybernetics" to propose a unified approach to the study of living organisms and machines (Rosenblueth, Wiener, and Bigelow 1943). More precisely, the authors focused on purposive adaptive behavior (in animals, including humans) and proposed the use of machines, built with the technology of their time, as "material models" for testing scientific hypotheses on behavior. They can be used for the following purposes:

• Corroboration or falsification: if the machine and the animal behave in the same or in a different way under the same external and internal circumstances, the hypothesis on the mechanism underlying the behavior is corroborated or falsified.

• Deciding between two competing hypotheses: if the behavior of the machine built according to the theoretical model M1 is more similar to the target biological behavior than the behavior generated by the robot built according to the theoretical model M2, the hypothesis corresponding to M1 is preferable.

• Generating new hypotheses on the functional structure of the biological system: for example, machines are used for science.

In fact, this concept was proposed earlier in some *proto-cybernetic* studies. In the early 1900s, Jacques Loeb and H. S. Jennings discussed mechanicism versus functionalism for studying the behavior of living organisms. They argued that if a machine is implemented on the basis of a theory of behavior, and it behaves according to what this theory allows to predict, then the proposed theory is reinforced. Again, machines are used for science.

More recently, this cybernetics approach has taken advantage of the progress of robotics technologies, bringing us to contemporary biorobotics. Now that we are back to the present, it is time to examine bioinspiration in practice and see the principles that we can learn from nature. Two of them, simplexity and embodied intelligence (introduced in section 1.2. as a motivation to soft robotics), are very general and encompass overall robot design and control, as explained in the next two sessions.

3.6   Simplexity

There is another word that you have to learn: *simplexity*. Alain Berthoz, a neuroscientist at College de France in Paris, wrote an enlightening book on this topic that is especially enlightening for roboticists (Berthoz 2012).

If you ask a roboticist why robots still appear clumsy in natural, unstructured environments, they will probably reply that a robot is a complex system, with many components like sensors and actuators, a complex control system, many connectors, and many lines of software code. The probability that some subsystem fails or that the overall robot behavior is different from expected is not zero. If you ask Berthoz the same question, he will reply that our robots do not work well because they are too simple. They are not complex enough for negotiating the complexity of the real world. Living beings are way more complex than our most sophisticated robots. Living beings have an indecent number of distributed sensors (receptors for sight, touch, hearing, and more); they have a huge number of degrees of freedom (joints) and actuators (muscles), to say nothing of the control system (brain and peripheral nervous system) and connections (nervous fibers). How is all this complexity coped with? Here comes the magic word just introduced: simplexity. Before explaining what it is, we need to understand what it is not. It is not simplification. It is not about simplifying our systems to simplify the way they work, and it is not about simplifying the environment to a structured one. Instead, it is about the principles and mechanisms that simplify the way living beings work, with complex bodily equipment, in complex environments. Such simplifying principles are what robotics should look at.

Definition 7 Simplexity Comprises a collection of solutions that can be observed in living organisms which, despite the complexity of the world in which they live, allows them to act and project the consequences of their actions into the future. It is not a matter of simplified model adoption but rather an approach to using simplifying principles. (Berthoz 2012)

We can then define simplexity as a number of simplifying principles that nature has put in place, through evolution, for guaranteeing survival. Despite the complexity of natural processes, simplexity principles enable living beings to process complex situations in a fast, elegant, and effective way by taking into account past experiences and by predicting future ones. Simplifying in this sense means inhibiting, selecting, connecting, anticipating. A simplexity principle that can inspire roboticists is anticipation and prediction. Brains have an excellent capability for prediction, based on what a person has learned from previous experiences. Anticipating future situations is an extremely powerful mechanism for simplifying the processing required to manage them.

A biological or robotic system is one that is built up from a sensory system that perceives external stimuli, a processing unit that receives such sensory inputs and plans actions, and a motor system that executes the movements back in the environment. And we usually set a precise sequential relation in the flow of information. This is how we described robot architectures in chapter 2 and how the human nervous system is also described, especially for educational purposes. However, our brain has more connections than expected between sensory and motor areas. Indeed, there is not a linear, sequential relation between them.

Going back to Berthoz, we should better note the following:

Perception is more than just the interpretation of sensory messages. Perception is constrained by action; it is an internal simulation of action. It is judgment and decision making, and it is anticipation of the consequences of action.
(Berthoz 2000, Chapter 1, p.9)

We can distill the main idea and use it in robotics. We can add predictions to sensory-motor architectures. As we learn from neuroscience, predictions are based on the current motor commands. So, the same motor commands issued for execution by the robot actuators are sent to our new predictor module. The predictor task is to generate an expected perception by anticipating what the sensory input will be after the robot executes the movement. This task cannot be accomplished without some previous experience of the movements, the sensory inputs, and the environment. As an example, when we grasp an object, we can generally predict the tactile image on our hand because we have previous experience of such a common action. All that previous knowledge is referred to as *internal models* in neuroscience. In robotics as well, we need to build models encoding this knowledge to enable the predictor to generate an expected perception. Once the predictor knows what movement is going to be executed and what perception usually follows that movement, then it can generate the expected perception. This latter action plays a very important role. We are not going to discard actual sensory inputs; rather, we want to compare them with the expected perception. If the prediction is good, we continue the current movement, and the sensory-motor loop is shortcut to a very short loop. In the case of a mismatch, we just execute the original, complete perception-action loop. Comparisons are generally fast, and matching can be assessed with similarly fast thresholding. Internal models can be built by the robot through offline learning phases.

See a complete robot architecture in figure 3.2. You may wish to compare it with figure 2.17. As you see, this architecture does not look simpler but more complex. However, there are several shortcuts that make the robot behavior simpler.

3.7   Embodied Intelligence

We tend to think that intelligence is shaped in our brain and our sensory-motor behavior is controlled by our central and peripheral nervous system. This is certainly true, but a modern view of intelligence tells us that our physical body has a role in all that (i.e., intelligence is embodied, as first introduced in section 1.2, as a grounding motivation for soft robotics). Movements and sensory-motor coordination are highly determined by our physical body, by its mechanical properties, by the way it interacts with the environment. Morphology and arrangement of muscles make a difference not only in the way we move, which is quite straightforward, but also in the way we control those movements, or better, the way we do not. Similarly, morphology and arrangement of receptors make a difference in how perception happens. The ultimate meaning is that morphology and physical interactions perform a part of the control job. Those processes performed by the body that otherwise would have to be performed by the brain are referred to as *morphological computation*. In other words, it is the part of computation done by the physical body for motor control or perception. A few examples can help clarify the concept of embodied intelligence.

If you would build two simple legs, connect them with a rotational joint, and put them on a slight slope, you would see them miraculously walk. You would have built a *passive walker*, and you may want to watch one of the many videos available online to see how it works. The system does not have any computing unit, and there is no control, not even actuators. It has only a body; its morphology allows for the walking movement, and the environment (i.e., the slight slope) provides the actuation forces, generated through gravity acceleration. The way the physical body interacts with this special environment, or ecological niche, determines the walking behavior. The passive walker is an extreme example of embodied intelligence. The ecological niche where it works (i.e., the slope) is quite limited, and the task that it can perform is just one (i.e., walking). It is, however, a very good example of the interplay between body, environment, and the tasks on which embodied intelligence is based.

If you need to be convinced that embodied intelligence takes place in more complex and helpful systems, we have to look at human walking. When we walk, we can easily adapt to a terrain that is not perfectly smooth, with no conscious brain effort. Of course, there are nervous mechanisms that do not involve the brain, like those happening in the spinal cord. But, as you can guess at this point, there is also a part of the job done by the legs themselves, which represent our embodied intelligence. Muscles and tendons are elastic, and our knee joint is compliant. Small adaptive movements are performed mechanically upon impact with the ground, without requiring neural control. The muscle-tendon system performs the morphological computation corresponding to the control of such adaptive movements.

Embodied intelligence is an extremely powerful simplifying principle that we can learn from nature. If properly designed, robots can take advantage of the physical interaction with the environment to perform the desired movements and reduce the computational burden on their control system. Referring to figure 3.2, embodied intelligence closes a very short loop between the environment and the robot mechanics. You just have to take into account two major things in robot design: robots are embodied, and design should always include the three components of body, task, and environment: in your design process, define the ecological niche, define the desired behavior and tasks, and then design the robot body.

Embodiment implies a number of properties that must be taken into account in robot design, not only to manage them but to take advantage of them. A very important one is that robots are subject to the laws of physics. While straightforward, this property introduces components that a robot may leverage. One is gravity. To obtain the intended movement, gravity needs to be compensated for, but it may also be leveraged—for example, for dynamic walking. Gravity provides a force a robot can use for moving, as in the passive walker example. Another important property of embodiment is that body movements generate perception (e.g., new sensory inputs come in when moving). This strong relation between perception and action was elaborated on in section 3.6. Also, body movements affect the environment. The importance of such interaction between the body and the environment is fundamental to embodied intelligence, as we have seen before. Embodied robots are dynamical systems and tend to settle into attractor states. Legged locomotion is again a good example. Quadrupeds tend to settle into a finite number of states, like walking, trotting, or galloping, even though the movements of their legs can be controlled independently, theoretically in a variety of locomotion patterns. The attractor states are determined by the physical properties of legs and body and their interaction with the environment (e.g., gravity). Finally, we mentioned that embodied robots perform morphological computation (i.e., a part of processing is performed by the body). More precisely, morphological computation can facilitate control, as in the case of the passive walker or the human knee compensating for uneven walking ground. And it can facilitate perception when receptor arrangements contribute to sensory processing. As an example, the insect compound eye receptors are not equally spaced, allowing them to detect an object's distance without the need for processing. A similar nonuniform arrangement of hearing receptors in some insects directly provides information on the sound source direction.

The simplifying principles implied by the properties of embodied robots blurs the boundaries between simplexity and embodied intelligence. Altogether, they represent a framework for more specific bioinspired principles beneficial to robotics, like those described in the following sections, which are examples picked from various suggestions taken from nature.

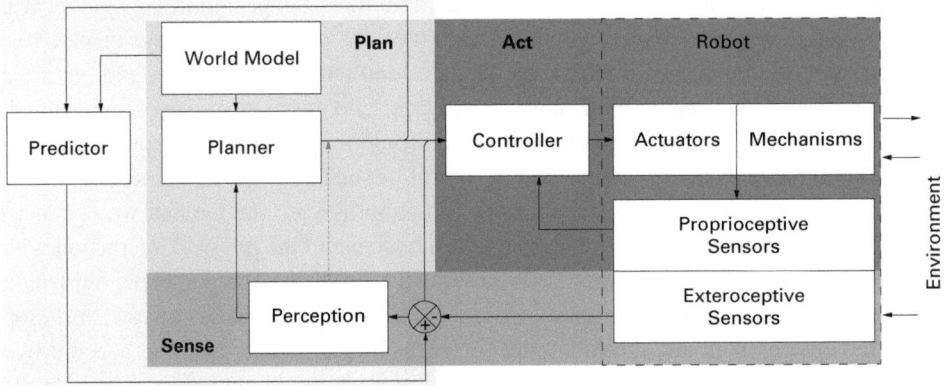

Figure 3.2
Bioinspired robot architecture. On top of a traditional sensory-motor control loop, implemented by either a hierarchical or a reactive architecture (see figures 2.17 and 2.18 in chapter 2), we add a prediction module that generates an expected perception, using the current motor command as generated by the planner, and an internal model. The expected perception is compared to the actual one coming from exteroceptive sensors, and if their difference is below a given threshold (i.e., the prediction is good enough), the current motor action continues, with no involvement of perception or planning. Otherwise, when conditions have changed and the prediction is no longer usable, the full loop is executed. Please also note the embodied intelligence in this bioinspired architecture, represented by the arrow from the environment to the robot mechanics. This figure shows how simplifying principles like prediction and embodied intelligence are implemented in modules that are added to the traditional robotics techniques. The system becomes more complex, but it is capable of using dramatic shortcuts whenever possible.

3.8   Bioinspired Principles in Control: Neurocontrollers

The way computer algorithms process information is very different from how it works in biological nervous systems. The basics in biological systems are neurons and how they are connected to form neural networks. Information flows through neurons in a very peculiar way: each neuron produces a spike in an electric signal that is received by all the neurons it is connected to; the spikes received by a neuron are summed up in a weighted way, and a spike is produced only if an activation threshold is reached. The network of connections determines the output from a given input. It generates the activation of muscles after the stimulation of sensorial neurons, or receptors, up to the generation of the cognitive processes in our brain. Connections and their weights can be updated, and this mechanism provides neural system with learning capabilities.

This strong principle observed in biology inspired the world of artificial neural networks. The principles that you can retrieve here are the basic unit of neuron, the weighted connections, the activation threshold, and the spike-like activation (i.e., a binary output). Of course, you find learning in artificial neural networks, which makes them beneficial as a computing paradigm and helpful in robotics as a means for robots to acquire learning abilities.

We recalled in chapter 2 the basics of robotics and robot control (i.e., the direct and inverse transformations between joint and Cartesian space and the diverse control loops that produce an intended robot motion). Neural networks can be used for most of those purposes by developing so-called neurocontrollers. Unlike traditional control approaches, neurocontrollers rely on learning for building the transformations between spaces. This means that a kinematic or dynamic robot model is not necessary but a learning phase is, where the neu-rocontroller can learn those transformations through actual robot motions. Notably, learning can be used for building the internal models needed in the predictive architectures outlined in section 3.6 (see also figure 3.2).

A full description of artificial neural networks is beyond the scope of this chapter, but you will learn how to use this approach when discussing soft robot control in chapter 6.

3.9 A Methodology for Biorobotics

After learning about bioinspiration and biomimetics and describing a few specific principles that we can take from nature to improve out robots, let us now review a general method that can help us in this process. Although it is clear that bioinspiration and biomimetics embrace a wide range of fields, we focus on robotics here, or biorobotics. The steps of the process are illustrated in figure 3.3. Please keep it on hand while going through the step detailed descriptions below.

Biological model
Our process starts with a biological system, of course. How it is chosen depends on the problem addressed on the technological side, or better, the ability we wish to provide our robot with. As we have seen, adhesion can be addressed by choosing an animal with an extraordinary dry adhesion ability, like the gecko. If we target biped walking, human beings are of course an excellent model. For flying robots, we may look at birds in general or a particular species of bird for specific flying abilities, like high-speed flight, energy-efficient gliding, or high maneuverability. Do not forget that plants are an excellent source of inspiration, too.

Principle
The process starts with the observation of the biological model, which can be done in many ways. First of all, an accurate literature survey, guaranteeing that the existing knowl-edge is properly taken into account, should teach the basics of the ability under study. Then, quantitative data should be extracted, as data are needed for the next engineering

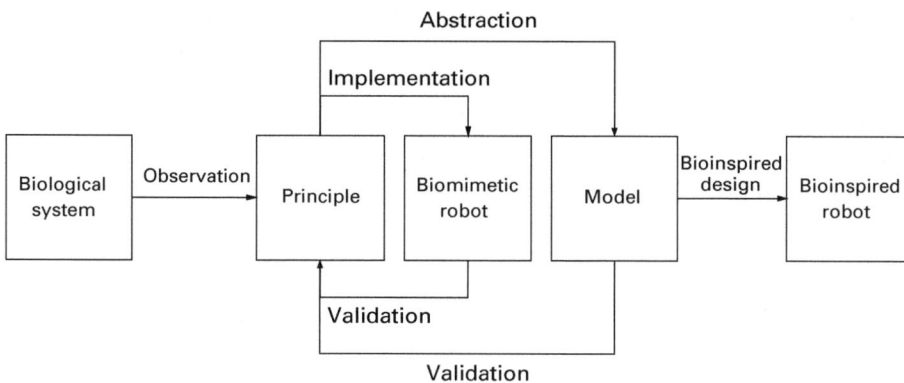

Figure 3.3
The biorobotics methodology.

work. If not available in literature, purposive experimental investigations should be done, preferably in collaboration with scientists, either in the wild or in lab settings. They range from the use of cameras for kinematic reconstructions or motion capture to imaging techniques like microscopy, echography, MRI, or other. As you see, this is very much dependent on the species observed and the target principle. The output of this first step of the process is a description, in as quantitative terms as possible, of the biological principle to use.

Biomimetic robot

The next step is the implementation of the principle into a biomimetic robot. Of course, this step has several intermediate steps, like a minimum level of abstraction of the principle into a mathematical description that can be used for designing the robot. However, it is important to stress here that early robotic prototypes used as proofs of concept can be a trial-and-error process. Indeed, prototypes can be used to test the principle experimentally. They play the role of physical models. They can help spot inaccuracies in the description of the principle and can help guide the choice of technological solutions afterward. It is helpful to keep the robot as similar as possible to the biological models in order to focus on the principle, without adding irrelevant variables to the scientific validation task. This loop can go on until the principle is validated.

Mathematical model

At this step, the principle is given a more formal mathematical description that represents an abstraction, going beyond the specific species considered. The modeling methods are given a purposive chapter in this book (Chapter 5) and are not treated here. What is important in this step of the process is that we are getting rid of the similarity with the biological system; at the same time, we have a tool for implementing the biological principle more deeply. This model can serve the purpose of validating the principle as well. After a few possible loops here, we are ready for the next and final step.

Bioinspired robot

We reach our target here. If everything went well, we would have a robot that embodies the principle we observed in nature. The robot does not need to be similar in appearance to the starting biological system. Indeed, this is not at all a criterion to judge whether a robot is bioinspired or not. Instead, our robot possesses the same principle and shows a similar ability or behavior, despite its shape or size. In fact, it can be scaled to the application it is designed for. A correct implementation of the principle abstraction that is encoded in the mathematical model guarantees the intended behavior.

You now are able to understand the slight difference between the meanings of biomimetic and bioinspired robots. Biomimetics is used here for the robots that mimic the principle to test for the purpose of validating it. It is the second direction of interaction between biology and engineering (i.e., the use of robots to study biological systems). Bioinspiration is used to indicate a higher level of abstraction of the biological principle, described mathematically and usable for designing robots that use the principle, which can be rescaled and reshaped to match the task the robots are intended to accomplish. It is the first direction of interaction, from biology to engineering. These definitions are not universally acknowledged in the literature, but they can be useful within the scope of this book.

3.10   Summary

You have learned that robots can be designed by taking inspiration from nature so they are able to operate in unstructured environment. The main purpose is to find those principles that nature has put in place for making the behavior of complex systems in complex environments simple and fast. Some inspiration can come from this chapter, which hopefully changed your views about robot perception, sensory-motor architectures, and controllers. And you now have a method for taking full advantage of biological inspiration.

Self-Assessment Questions

1. Can you identify a bioinspired object in your daily life?

2. What makes a robot bioinspired?

3. Find examples of embodied intelligence in movement and perception in nature.

4. What are the main differences between algorithmic processing and biological computation?

5. Name and explain one biological principle of your choice that is interesting for robotics. Test yourself in applying the main steps of the biorobotics methodology.

4 Soft Robotics Technologies

Chapter Objectives

- To recall the basics of material mechanical properties
- To list a few typical materials for soft robotics
- To learn the working principles and mathematical descriptions of a selection of the actuators and sensors used in soft robotics
- To learn approaches and technologies for stiffening
- To list a few approaches for designing deformable structures

4.1 Overview

This chapter focuses on how to build a soft robot. Chapter 1 has outlined a few possible approaches that are consolidated in the field, while research is going on with exploring more materials and actuation and sensing technologies. Here, we are going to examine those approaches more in depth, describing the technologies involved, in terms of working principles and relevant mathematical descriptions. We are going to consider materials, actuation technologies, and sensing technologies. Please keep in mind, though, that it is a stretch for educational purposes. In soft robotics, boundaries between materials, actuation, and sensing are very blurred, thanks to smart materials that can move and sense.

As a first step, we recall some basics of mechanical properties of materials, and then we review some of the materials most commonly used in soft robotics, like rubbers, silicones, and hydrogels. We then dive into a selection of actuation technologies that are either coupled with soft materials or are smart materials capable of movement in their own right. We do the same for sensing, discovering that some of these technologies are similar. In soft robotics, stiffening is a relevant aspect that complements actuation. It deserves a dedicated description of the possible approaches and technologies.

As an attentive reader, you may think that we are forgetting a part of the soft robot definition.[1] What has been described so far refers to the first part ("soft robots/devices . . . relying on inherent compliance")—that is, meaning built with soft materials. Actually, we give account to the second part of the definition—"soft robots/devices . . . relying on structural compliance"—by mentioning the possible alternative approaches based on tensegrity, meta-materials, origami, and kirigami. We dive less deep into those concepts, though.

4.2   Materials for Soft Robotics

The title of this section is quite ambitious. We cannot review all possible materials usable in soft robotics. It is however helpful to recall the main properties of materials for purposes of having the right tools to choose the best-suited material for our soft robot design. In any case, there are some materials already widely used in soft robotics, and it is worth knowing their main properties.

4.2.1   Material Properties

In soft robotics, we are especially interested in the mechanical behavior of a material. We are interested in the way it deforms in response to a stimulus. Before defining the deformation and the stimulus more precisely, please consider the general behavior of a material under tensile or compression loading conditions. Then we can define our terms less generically, and we can represent them graphically for this case (see figure 4.1):

• **Stress**. This is the mechanical stimulus. It has the form of a pressure (i.e., a force applied on a surface). The force can be a compression or a tension. In a cylindrical sample, it is the pressure applied on the circular cross-section.

$$stress \; \sigma = \frac{F}{A} \tag{4.1}$$

• **Strain**. This is the deformation. It is the change of a geometrical parameter with respect to its original value. It is then a relative change. In a cylindrical sample, it is the relative length change.

$$strain \; \varepsilon = \frac{\Delta L}{L} \tag{4.2}$$

The mechanical behavior of a material is described by stress-strain curves. A general stress-strain curves is given in figure 4.2. Deformations can be *elastic* (i.e., reversible when the load is removed) or *plastic* (i.e., permanent, maintained after the load removal). In general, a material can deform in an elastic way up to a given threshold called the *elastic limit*. Over that threshold, the deformation enters the plastic regime and is no longer fully reversible. Another, higher threshold sets the material highest resistance (*ultimate strength*) before the complete failure of the material (*fracture*).

We now have all the basic ingredients for defining important parameters that describe the mechanical behavior of materials:

• **Young's modulus *E***. It is the ratio between the stress applied to the material and the corresponding strain (i.e., its deformation) when the load is applied along the same axis of deformation (as in figure 4.2). Please note that since strain is a dimensionless quantity, the Young's modulus has the same unit as pressure and is expressed in Pascals (*Pa*) in SI.[2]

$$E = \frac{\sigma}{\varepsilon} \tag{4.3}$$

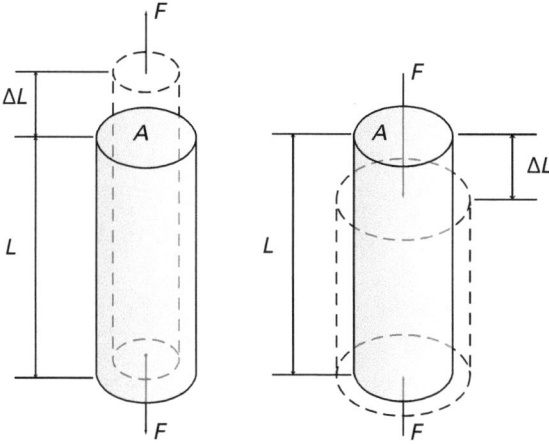

Figure 4.1
Stress and strain on a cylindrical material sample. (a) Tensile load. (b) Compression load.

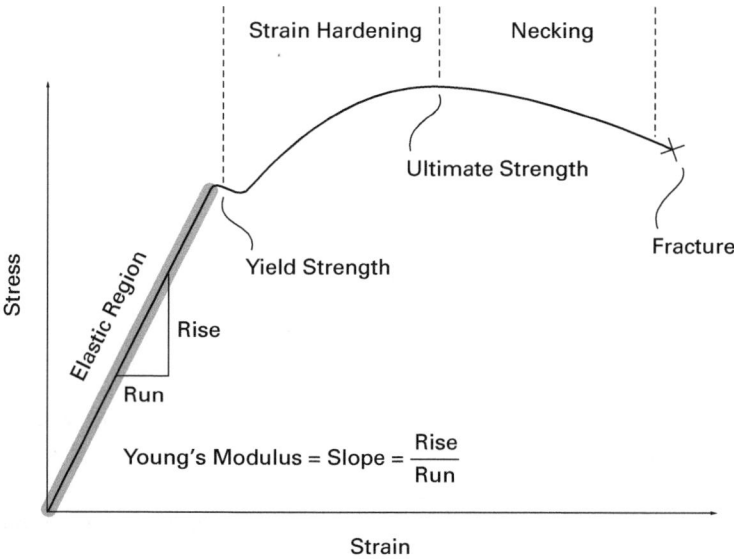

Figure 4.2
Typical stress-strain curve of a material, showing elastic region, and fracture.

• **Shear modulus G.** This is the ratio between shear stress and shear strain—that is, the deformation under a lateral force (see figure 4.3).

$$G = \frac{\frac{F}{A}}{\frac{\Delta x}{l}} \qquad (4.4)$$

• **Bulk modulus K.** This is a measure of the deformability of the material in all directions, or volumetric elasticity, considering the volumetric strain under a volumetric stress (i.e., a stress uniformly applied in all directions). You may see it as an extension of the Young's

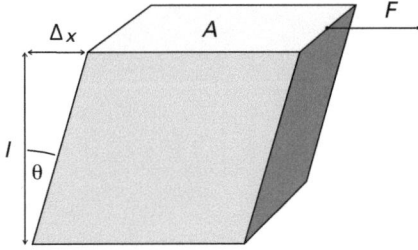

Figure 4.3
Illustration of shear stress and shear strain, with relevant parameters.

modulus to three dimensions, and it is the inverse of compressibility. It is defined as the ratio between the pressure (P) variation resulting from the volume (V) variation.

$$K = -V \frac{\mathrm{d}P}{\mathrm{d}V} \tag{4.5}$$

• **Poisson's ratio v.** This is the variation of the cross area with respect to the strain (i.e., the length variation in our cylindrical sample). In general, if the cross area increases the length decreases, and vice versa. Please note that its formula below includes the negative sign in front of the ratio to compensate for the negative sign deriving from that. You may think this detail is not especially interesting, but please keep it in mind, because it will become relevant for the metamaterials described in section 4.6.

$$v = -\frac{\dfrac{\delta A}{A_0}}{\varepsilon} \tag{4.6}$$

• **Ductility** or **elongation %.** This is the ability to deform before fracture under tensile load. It is expressed as percentage elongation at fracture—namely, $\varepsilon_{fracture} \times 100$.

• **Stiffness k** and **compliance k^{-1}.** They are inverse to each other. They are a measure of the resistance offered to a force F, or inversely the deformation produced by the force, respectively. The deformation is expressed by a displacement Δx. The *stiffness* parameter is commonly used for springs. The directly proportional relation between the deformation or displacement and the applied force was discovered by the British scientist Robert Hooke in 1660, hence it is well-known as Hooke's law:

$$F = k\Delta x \tag{4.7}$$

We can derive the *stiffness* parameter k as follows:

$$k = \frac{F}{\Delta x} \tag{4.8}$$

While E is an intrinsic material property, k is also depending on geometrical features. For the case reported in figure 4.1, the stiffness k can be expressed as a function of E:

$$k = E\frac{A}{L} \tag{4.9}$$

as well as the compliance k^{-1}:

$$k^{-1} = \frac{L}{EA} \tag{4.10}$$

4.2.2 Soft Materials for Soft Robots

We now have the tools for describing a material, using the main parameters that describe its mechanical behavior. If we consider the Young's modulus only, we can already make a good ranking of soft materials. We can set a threshold at 10^9 Pa (1 GPa), thus including rubbers, silicones, and hydrogels within soft materials (see figure 4.4).

• **Rubber**. It is a polymer that can stretch and shrink. Rubber is used for a wide range of objects used daily, from gloves to car tires. It has been produced for decades from natural sources (from *Hevea brasiliensis* trees), and it can be synthesized on an industrial scale today. Its Young's modulus is 5×10^7 Pa, and the maximum stress it can withstand is 5×10^6 Pa.

• **Silicone**. Figure 4.5 shows the example of a stress-strain curve for a silicone, obtained with tensile mechanical test. We observe different values for the Young's modulus in different parts of the plot (i.e., at different levels of strain). With strain as low as 25 to 75 percent, the Young's modulus is close to 2.5×10^5 Pa. At higher strains of 175 to 225 percent, the Young's modulus reaches 9×10^5 Pa. The maximum stress that this material can withstand is 2×10^6 Pa. The ultimate strain reaches 325 percent. In general, the silicone materials used in soft robotics reach hundreds of percent of strain. Synthetic silicone composites show relatively good resilience compared to, for example, natural rubber.

• **Hydrogels**. They are multiphase composite materials consisting of an aqueous matrix reinforced by a solid polymer network. Hydrogels come from several natural sources, like collagen, gelatin, agar, and alginate hydrogels. They can also be synthesized from engineered polymers, like poly(ethylene oxide) (PEO), poly(2-hydroxyethyl methacrylate) (poly-HEMA), poly(-vinyl alcohol) (PVA), and poly(-acrylamide) (PAAm). The Young's modulus of natural and synthetic hydrogels may be significantly different—for example, 8×10^3 Pa and 24×10^3 Pa. Hydrogels can retain large amounts of water within their intermolecular space because of strong hydrophilicity of the polymer chains and large porosity. As such,

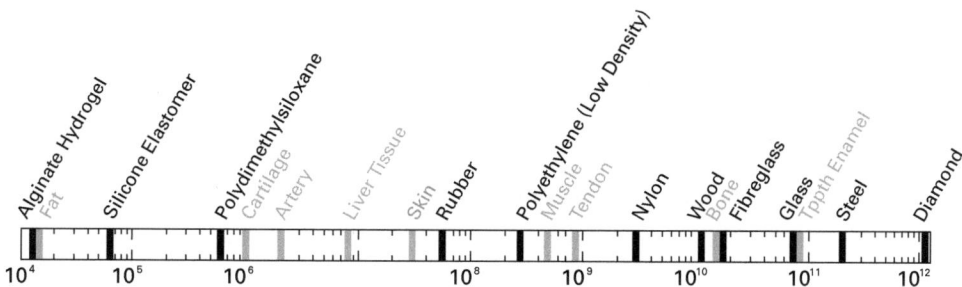

Figure 4.4
Young's modulus of main biological tissues and a few materials (reproduced with permission from Rus and Tolley 2015). Setting a threshold at 10^9 Pa identifies some rubbers (between 10^7 and 10^8 Pa), silicones (between 10^5 and 10^7 Pa), and hydrogels (between 10^4 and 10^5 Pa) as soft materials.

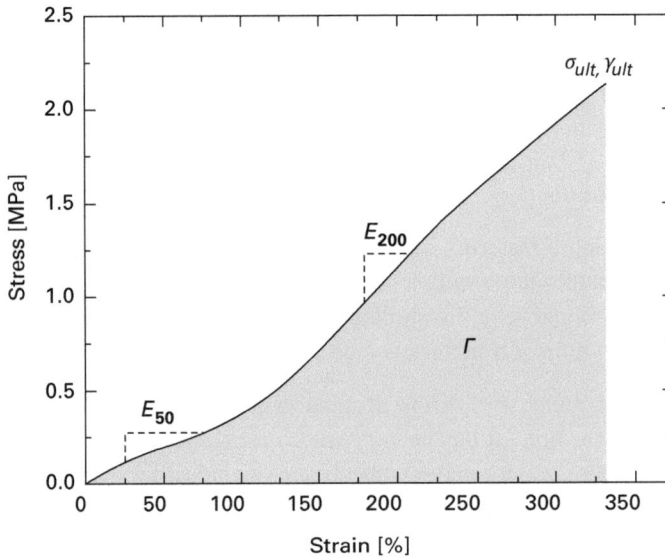

Figure 4.5
Example of a tensile loading curve for Mold Max 10 (made by Smooth-On, Inc.), a tin-cured silicone composite material. Reproduced with permission from Polygerinos et al. (2017), figure 1.

hydrogels can undergo significant swelling in water, from 10 percent to 1,000 times their dry weight.

4.3 Actuation Technologies for Soft Robotics

Actuation is what we need to make our soft robot move. An actuator is the part of a mechatronic system that produces a physical change by generating a force, a torque, and ultimately motion. More specifically, an actuator converts some form of energy into motion. It is a transducer, in this respect. The input energy, for our robotics purposes, is generally electrical (a voltage or a current) and is converted into mechanical, sometimes through fluidic (either hydraulic or pneumatic) mechanisms.

Among the properties of an actuator, the following should be considered when designing or choosing a robot actuation system:

· **Continuous power output**. The maximum force/torque in a time unit attainable continuously

· **Range of motion**. The range of linear/rotary motion

· **Resolution**. The minimum increment of force/torque attainable

· **Accuracy**. The ability to produce an output that closely reflects the input command

· **Peak force/torque**. The force/torque at which the actuator stalls

· **Speed characteristics**. Force/torque versus speed relationship

· **No load speed**. Typical operating speed/velocity with no external load

· **Power requirement**. Type of power (AC or DC), number of phases, voltage level, and current capacity

· **Efficiency**. The ability to convert input energy into useful mechanical motion with minimal energy loss, which is affected by factors like friction, heat generation, and mechanical resistance

Without losing generality, let us anchor to the soft arm depicted in figure 1.1(a), for the rest of this chapter. It will be easier to go through different actuation technologies and see how they can produce the same arm movement. For the same reasons, let us focus on a bending motion, which is a very typical soft robot movement. In a rod-like soft arm like the one in figure 1.1(a), bending on a side can be obtained by reducing the length on the same side or increasing the length on the opposite side of the arm. We can then describe how different actuation technologies can produce either shortening or elongation on one side. All you are going to learn is nonetheless usable in the variety of other soft robot designs and morphologies.

4.3.1 Tendon-Driven Actuation of Soft Robots

We can suppose the soft arm of figure 1.1(a) is made of silicone or another soft material, as discussed in section 4.2.2. The easiest way to obtain bending by shortening one side of the arm is to embed a cable in the soft material and pull it to act as a tendon. Its action would be distributed along the soft material. A simple way to pull the cable is to attach it to a motor that can be conveniently located outside the arm itself (see figure 4.6).

The parameters that come into play and their relations are described by the following equation:

$$q = \frac{T(aRk - 1)}{E\pi R^2} \tag{4.11}$$

where q is the longitudinal strain, T is the cable tension provided by the external motor, and E is the Young's modulus of the soft arm material. We assume that the distance between the cable and the midline is proportional to the radius of the section R, and we use a scalar value a that indicates the position of the cable with respect to the section so that aR is the distance of the cable from the midline. Finally, k represents the curvature.

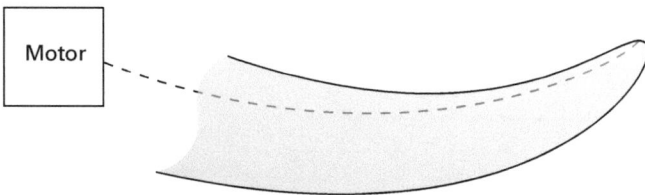

Figure 4.6
Bending by a tendon. A cable is embedded inside the soft arm material, asymmetrically with respect to the central axis, so that it acts as a tendon on one side of the soft arm. When pulled by an external motor, it shortens one side of the soft arm, generating a bending movement in the same direction.

4.3.2   Fluidic Actuation in Soft Robotics

Fluidic actuation is the use of fluid pressure for producing an actuation torque and motion. In soft robotics, fluidic actuation is coupled with a flexible structure, so the term flexible fluidic actuator (FFA) is used in this field. Both liquids and gases are used, and in most cases, they are simply water and air, respectively. The underlying principle for actuation is that a soft-bodied chamber is deformed when the pressure of the fluid that it contains changes. The pressure change is generally an increase, and the corresponding deformation is a chamber expansion, but the use of negative pressure is also possible. The chamber expansion can be guided, by design, toward predefined deformations. In our bending case, we may wish to have an elongation on one side of the soft arm so that it bends in the opposite direction, as depicted in figure 4.7.

Actually, flexible fluidic actuators can both elongate and shorten—that is, they can play the role of either extensors or contractors. Let us examine a few specific cases.

In chapter 1, when recalling a bit of history of soft robotics, McKibben actuators stood up as one of the first, ante-litteram pneumatic actuators for today's soft robotics. They consist of a hollow rubber tube, generally cylindrical, contained inside a woven sheath, also cylindrical. When pressurized, the tube itself would tend to expand in all directions, like a balloon. The external sheath constrains the radial expansion and guides it to become a longitudinal deformation. A good question here is whether such deformation is an extension or a contraction. Actually, it can be both, and it depends on the angle that the fibers of the external sheath form with respect to the longitudinal axis. There is a critical angle where neither of the two happen. Instead, with a larger fiber angle the tube extends, while with a smaller angle the tube contracts. We see why in the mathematical relations that bind the geometric parameters involved and described in figure 4.8. Consider one fiber only, wrapped around the cylindrical tube with n turns. Unwrap the n turns to visualize a triangle, as shown in figure 4.8. The base of the triangle will be n times the circumference of the cylinder—that is, $n\pi D$, with D being the tube diameter. The triangle height is the same as the cylinder length, L. The length L and the diameter D relate to the fiber angle θ as follows:

$$L = b \cos \theta \tag{4.12}$$

$$D = \frac{b \sin \theta}{n\pi} \tag{4.13}$$

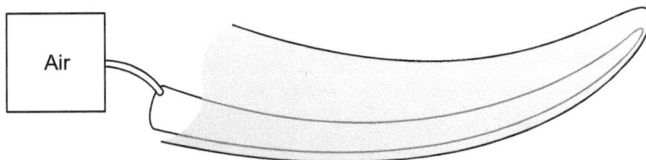

Figure 4.7
Bending by a flexible fluidic actuator. A cavity in the soft arm material, asymmetrical with respect to the central axis, deforms into an extension when the fluid pressure increases and elongates one side of the soft arm, generating a bending movement in the opposite direction.

where b is the third triangle side (in figure 4.8) and can be easily calculated as follows:

$$b = \sqrt{L^2 + D^2 n^2 \pi^2} \tag{4.14}$$

At this point, we can express the volume of the cylinder as a function of θ:

$$V = \pi \left(\frac{D}{2}\right)^2 L = \frac{b^3 \cos(\theta)\sin^2(\theta)}{4n^2\pi} \tag{4.15}$$

This function shows a maximum, meaning that the volume increases until a specific value of θ and decreases afterward, with a growing θ. We can find the corresponding θ value as the point where the derivative equals 0:

$$\frac{dV}{d\theta} = \frac{b^3}{4n^2\pi}(2\sin\theta\cos^2(\theta) - \sin^3(\theta)) = 0 \tag{4.16}$$

$$\theta_{max} = 54.7° \tag{4.17}$$

The axial tension F can also be expressed as a function of θ and of the relative pressure P', by using the virtual work principle:

$$P'\mathrm{d}V + F\mathrm{d}L = 0 \tag{4.18}$$

$$F = -P'\frac{\mathrm{d}V}{\mathrm{d}L} = \frac{P'b^2(3\cos^2(\theta) - 1)}{4n^2\pi} \tag{4.19}$$

An alternative to the woven sheath of McKibben actuators for obtaining extensions/contractions are other forms of longitudinal or transverse constraints for the walls of the pneumatic chamber. An example is the use of fibers embedded in the soft material of the chamber or corrugated surfaces, as shown in figures 4.9(a) and 4.9(b). For instance, longitudinal fibers embedded in the walls of a cylindrical pneumatic chamber would allow transverse expansion but would constrain the overall length, generating a contraction. Likewise,

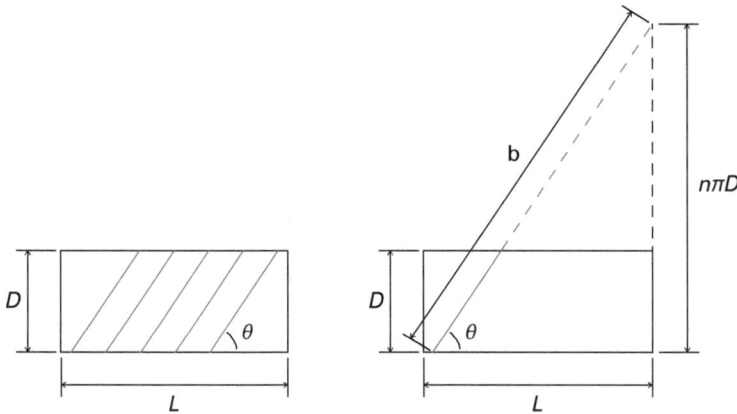

Figure 4.8
The geometrical parameters describing a McKibben sheath.

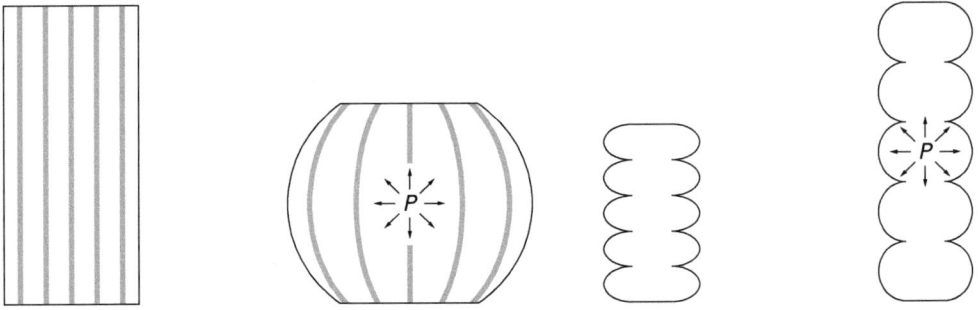

Figure 4.9
The use of longitudinal fibers or corrugated membranes for a cylindrical pneumatic chamber (a) contraction or (b) extension, respectively.

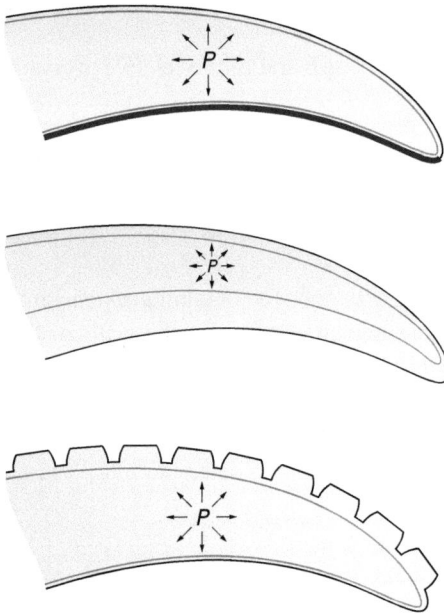

Figure 4.10
Strategies for driving FFA deformations: (a) a stiffer material on the bending side, (b) eccentric chamber on the side opposed to the bending side, and (c) patterning on external or external surfaces, opposite to the bending side.

but oppositely, a transversally corrugated surface would offer preferential direction of expansion along the longitudinal axis, at a lower pressure than the one required for transverse expansion.

Going back to our bending movement of figure 4.7, instead of using contractions or extensions as seen above, we can generate a bending directly. We can apply the same approach of adding constraints in the soft chamber and offer preferential deformations toward a bending this time. Figure 4.10 summarizes some of them, still visualized through our soft robot arm of figure 1.1(a). As a first idea, we may use a stiffer material on one side of the soft arm so that pressure increase would inflate the softer part first. That would generate a bending in the direction of the stiffer material, as in figure 4.10(a). An eccentric chamber on a side of the arm generates a bending in the opposite direction, as shown in figure 4.10(b). Adding a transverse pattern on the internal or external surface of one side of the chamber would facilitate an extension on that side, bending the arm in the opposite direction, as in figure 4.10(c).

It appears clear that a variety of strategies can be used in the design of specific soft robots for given applications. This is still very much a creative process, with no structured methods or theories.

4.3.3 Electro-Active Polymers as Soft Robot Actuators

If you take a layer of a soft material and put two conductive electrodes on the two opposite surfaces, you see the material squeezing when you apply a high voltage between the two electrodes, as they tend to attract to each other (see figure 4.11, top). This is the basic working principle of a kind of electro-active polymer (EAP). Actually, this name encompasses a large family of devices differing by operating conditions (i.e., wet or dry) and by working principles, from osmotic pressure and ion exchange in wet ones to piezoelectricity and electrostatic attraction in dry ones. The case described above refers to dry, dielectric EAP (DEA).

The electrostatic attraction of two electrodes in the direction of the applied field squeezes the soft layer in the thickness direction and expands it in the directions perpendicular to the field. This is given by the Maxwell stress effect. The voltage needed for the phenomenon to occur is generally very high, but we can dimension the final actuation effect by referring to a few mathematical relations between the electric field, the main geometrical parameters, and the properties of the material used.

Our EAP device is a capacitor, and the electrostatic energy it stores is given as follows:

$$U = \frac{1}{2}CV^2 \tag{4.20}$$

where C is its capacitance and V is the applied voltage. The capacitance C depends on geometric parameters and material properties as follows:

$$C = \varepsilon \frac{A}{t} \tag{4.21}$$

where $\varepsilon = \varepsilon_0 \varepsilon_r$ is the product of the material dielectric constant and relative permittivity, respectively, A is the area, and t is the thickness—that is, the distance between the two electrodes. We can then write the energy as follows:

$$U = \frac{1}{2}\varepsilon\frac{A}{t}V^2 \tag{4.22}$$

The force F and the stress S between the two electrodes and normal to them are thus:

$$F = -\frac{dU}{dt} = \varepsilon\frac{A}{t^2}V^2 \tag{4.23}$$

$$S = \frac{F}{A} = \frac{\varepsilon V^2}{t^2} \tag{4.24}$$

The geometric parameters then play a significant role in the final stroke that can be obtained in an EAP-based actuator, in addition to voltage and material properties.

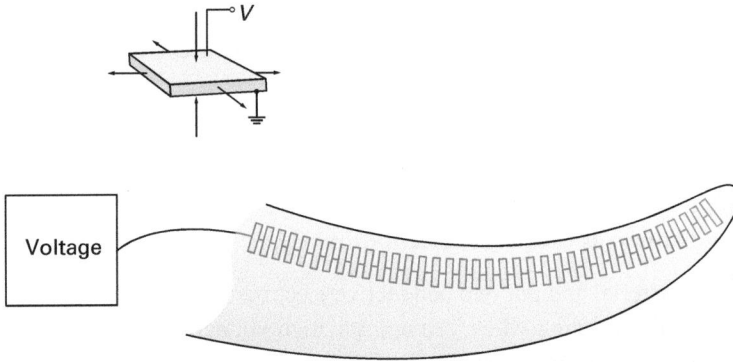

Figure 4.11
Bending by EAP. A stack of EAP on one side reduces its length and generates a bending on the same side.

This basic principle can be implemented in diverse geometries, including stacks that maximize contraction. Figure 4.11 illustrates how the contraction of a stack of EAP can reduce the length on one side of our soft arm and make it bend on that side.

4.3.4   Shape-Memory Alloys and Polymers

With EAP, we entered the world of smart materials, which are materials that can move and sense and thus provide interesting functionality for soft robotics. Shape-memory materials are another large class of smart materials relevant to soft robot actuation. As their name suggests, they have the property of recovering a preset shape when their temperature increases. More precisely, the shape-memory effect (SME) is the ability to recover a shape, even after an apparently permanent, plastic deformation occurred at a relatively low temperature, by heating to a temperature above a characteristic threshold. This effect is acquired, meaning that thermomechanical treatments are needed. In the one-way memory effect, the shape is changed mechanically at low temperature and goes back to the previous, undeformed state when the temperature rises. In two-way memory effect, the shape changes with temperature rise and goes back to the previous shape when the temperature drops. Depending on the material, the working principle is different but the effect is the same.

Shape-memory alloys (SMAs), like NiTiNOL,[3] do the shape recovering thanks to martensitic transformations. These transformations are thermally activated and reversible. They occur when a transition between austenitic and martensitic phases causes a structure change, given by small, coordinated shifts of the atomic positions, without plasticity. At low temperature, the alloy is in a martensitic phase and can be easily deformed mechanically. At a higher temperature, the alloy transitions to an austenitic phase and recovers the preset martensitic shape. When cooled down, it maintains the same shape until deformed again. Figure 4.12 illustrates the structure changes in the martensite-austenite transitions.

Stretched NiTiNOL wires shorten when heated. The application of a current is a simple way to heat them. The thermomechanical behavior of such an SMA wire can be described in terms of stress σ, total strain ε_t and absolute temperature T. We consider the Young's moduli for austenitic and martensitic phases, E_M and E_A, respectively. The inelastic transformation strain ε_i and the martensitic volume fraction $\xi_M = \dfrac{V_M}{V_{tot}}$ (ratio of volume in martensitic

Austenite

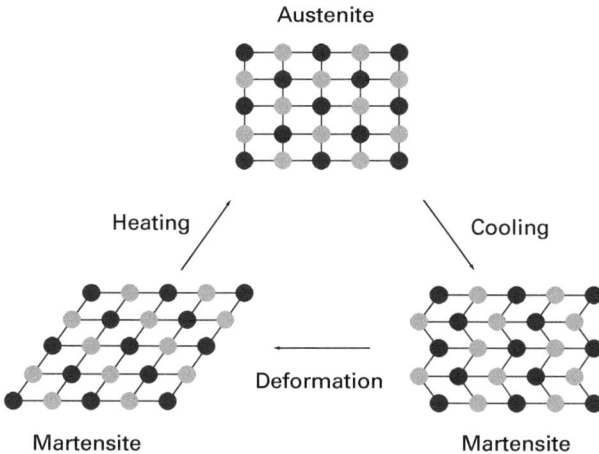

Heating / Cooling

Deformation

Martensite   Martensite

Figure 4.12
In martensitic phase, the alloy can be deformed. Above a characteristic temperature, the alloy enters the austenitic phase, showing a greater crystallographic symmetry.

phase V_M and total volume V_{tot}) also come into play, as well as the ambient temperature T_0 and the thermal expansion coefficient α_T. We can sum up the elastic, thermal, and inelastic contributions to express the total strain as follows:

$$\varepsilon_t = \left[\frac{1}{E_A} + \xi_M \left(\frac{1}{E_M} - \frac{1}{E_A} \right) \right] \sigma + \alpha_T \left(T - T_0 \right) + \varepsilon_i \tag{4.25}$$

When describing the thermomechanical behavior, we are interested in relating it to the electrical power that we provide to the SMA wire, i^2R, given by the current i injected in the wire and its resistance R. This is reduced by thermal inertia phenomena, like cooling, described by the thermal exchange coefficient h, the thermal exchange surface A, and the temperature T as $hA(T(t) - T_0)$. The power absorbed and used for increasing the SMA temperature is shown as follows:

$$mcV \frac{dT(t)}{dt} + mV\Delta H \frac{d\xi_M}{dt} = Ri^2(t) - hA(T(t) - T_0)$$

where m is the mass density, c is the specific heat, V is the volume, and ΔH is the latent heat for the transformation.

NiTiNOL wires can be shaped in springs that contract when heated. Figure 4.13 illustrates a possible use of a SMA spring in our soft robot arm for our usual bending movement. As mentioned, a heat treatment is needed for memorization of the spring shape so that it contracts. The process involves these main steps:

1. coiling the NiTiNOL wire in the spring shape while holding the part in a fixture during treatment

2. maintaining the spring at 450°C for 30 minutes under flowing nitrogen or other inert gases (to prevent oxidation), which is the shape memorization

3. water quenching the spring after this time has elapsed.

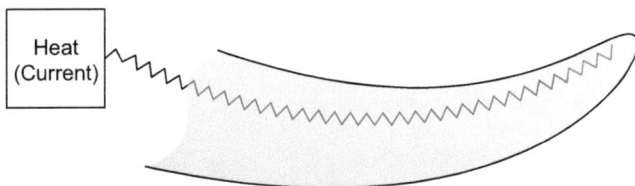

Figure 4.13
Bending by SMA. A spring made of SMA contracts when heated and reduces the length of the arm side where it is placed, generating a bending movement on the same side.

Shape-memory polymers (SMPs) differ from SMAs in their shape-memory effect. An SMP consists of a glass transition or melting transition from a hard to a soft phase. An SMP can retain two or sometimes three shapes, and the transition between those is induced by temperature, an electric or magnetic field, light, or a solution.

4.4   Stiffening Technologies for Soft Robotics

In soft robots, actuation is not the only important factor. Movement is key, of course, but the soft robot stiffness is an equally important aspect, making the difference in the robot's ability to accomplish tasks, like manipulation or locomotion, which require the application of forces on the external environment. While keeping the deformability of our soft robot, we may wish to add mechanisms for increasing the stiffness, when necessary, in a controllable way.[4]

A way to increase the stiffness with no shape change is suggested by nature. In most musculoskeletal systems, including the human one, a joint is attached to a pair of agonist-antagonist muscles that make it rotate in opposite directions. When contracted at the same time, the joint cannot rotate, while its stiffness increases. The McKibben actuators described in section 4.3.2 are often used in this configuration to actuate revolute joints. The same principle can be used with soft robots with no articulated structure. For instance, the antagonistic configuration can be given by actuators arranged longitudinally and transversally, as shown in figure 4.14, suggested by natural models like the octopus arm and the elephant trunk, coupled with analogous isovolumetric constraints. Or, actuators with opposite functions, like contractors and extensors, can be paired to work in discordant way and then increase the stiffness.

Stiffening mechanisms can also be added to the actuation system as separate and independent components. A physical phenomenon that can be used for stiffness variation is jamming transition. You may experience it every day when you buy vacuum-packed coffee or rice or similar products. The pack is very rigid indeed, but it becomes floppy as soon as you cut it and air can flow in. You just caused a jamming transition, in fact. More precisely, the jamming transition happens when the density increases and the viscosity increases as a consequence. It is fully reversible, as you see when opening a vacuum-sealed pack. It happens with some materials at the mesoscale, like granular materials, layers, fibers, and even purposively designed chained elements. With the density increase, the particles tend to stick to each other and even interlock. The density that triggers jamming is given by several factors, like morphology and deformability of the particles as well as frictional forces. A

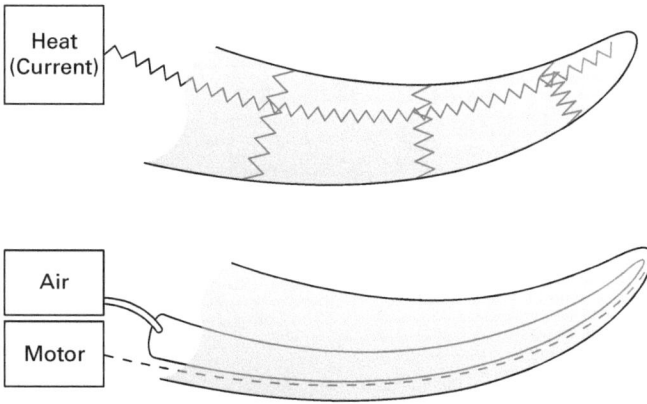

Figure 4.14
Stiffening by antagonistic actuators. (a) The longitudinal SMA spring tends to increase the cross-section when shortening and bending the arm; the transverse SMA springs contrast this action when contracting. Overall, the stiffness increases. (b) The pneumatic chamber and the cable have opposite actions, since the first elongates its side of the arm while the latter shortens it. While both can generate a bending, their combination increases the stiffness by contrasting forces.

complete theoretical description of the jamming phenomenon is still elusive. Basic relations can be devised from an ideal case of spherical particles in a confined space. Their density ϕ is the sum of their volumes V_P divided by the overall volume V:

$$\phi = \frac{NV_p}{V} \tag{4.27}$$

When ϕ reaches the jamming value, the particles come into contact with each other and cannot move anymore. The way they are in contact is chaotic, though, and we can just consider an ideal case, with an angle θ describing a regular arrangement, as shown in figure 4.15. Considering the 2D case, we can calculate the density with areas instead of volumes. In the simple case of four particles of radius r, we have four times each area of πr^2 and a total area of $16r^2 \sin 2\theta$ (it is the area of the parallelogram with base $4r$ and height $4r \sin 2\theta$). The density is then represented as follows:

$$\phi = \frac{4\pi r^2}{16r^2 \sin 2\theta} = \frac{\pi}{4 \sin 2\theta} \tag{4.28}$$

The angle θ can vary between 45°, when the particles are perfectly layered, and 60°. So, considering an average value of 52.5°, we have a jamming density $\phi \cong 0.813$, while averaging ϕ, we find $\phi \cong 0.824$. Extending to 3D, $\phi \cong 0.639$.

Returning to soft robotics, granular jamming can be used to create stiffening chambers that are filled with a granular material and connected to a vacuum pump. The stiffening chambers are added to the soft robot and activated once the robot has reached the position where stiffening is required. This is a perfect mechanism for stiffening with no shape change. Figure 4.16 illustrates the concept for our bending soft arm, which can bend and then stiffen in the same position with an internal granular jamming chamber.

Which material and which membrane work the best is still elusive, in theoretical terms. Experimental studies compared a few materials and outline the importance of design

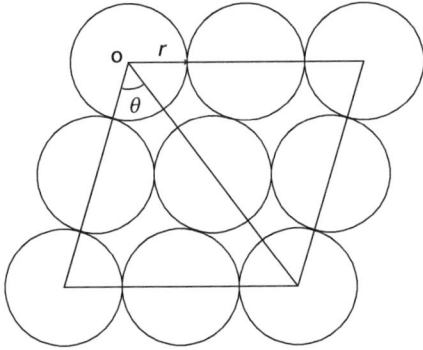

Figure 4.15
The ideal case of spherical particles inside a confined space. The angle θ is set by the line connecting the centers of the particles. The total space is shown by the outer boundary.

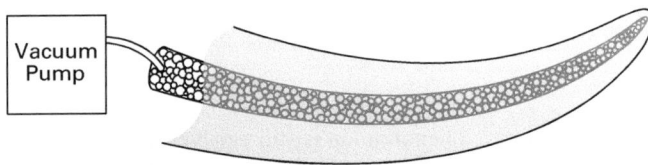

Figure 4.16
Stiffening by granular jamming. An internal chamber filled with a granular material stiffens the arm in its position when air is removed with a vacuum pump.

parameters such as the size, deformability, shape, and texture of the granular material and the thickness, elasticity, and texture of the membrane.

4.5   Sensing Technologies for Soft Robotics

In chapter 2, we have seen how important sensors are for robots. We outlined two major needs: having a feedback on the current robot position, for control purposes, and perceiving the external environment, for planning the robot behavior. With soft robots, we may have the same needs, but several things dramatically change. Above all, obtaining feedback on the current robot position requires a different approach than using the position sensors described in chapter 2. Those sensors are all supposed to be used on joints that connect two rigid links or on motors that are the typical actuators. In soft robots, we have neither of them. Instead, both the current position and the effect of external forces generate a deformation, which is what we need to sense in soft robots. Furthermore, there are some additional requirements for sensors to be embedded in soft robots, like being as compliant as the robot in order not to introduce rigidity, and being extensible enough to follow robot extensions, thus not creating stress concentrations that would affect the robot deformations. Some of the technologies seen in chapter 2 are also used for soft sensors, with some differences.

4.5.1   Resistive Soft Sensors

We introduced the piezoresistive effect in chapter 2, and we know that conductive materials change their electrical resistance with strain. Unfortunately, most soft materials are not conductive or tend to lose their conductivity at high strain. Nevertheless, a few solutions are available, as well as a few alternative approaches to use in soft robotics, that still rely on resistive properties.

Soft materials like those listed in section 4.2.2 can be made conductive. A possible way is to embed conductive particles into them. They can be metal nanoparticles, carbon nanotubes, carbon black, or graphene. Such additives may add stiffness to the material, though. Effective trade-offs are possible that give enough conductivity at an acceptable stiffness increase.

A conductive polymer placed between two overlapping arrays of electrodes tends to squeeze when pressure is applied. The electrical resistance between the two electrodes thus changes and can be measured, giving a measure of the pressure applied. Piezoresistive inks are used for this purpose, in the force sensing resistor (FSR) sensors cited in chapter 2.

Soft materials can also be embedded with channels where a conductive fluid is inserted. That can be an ionic fluid or a liquid metal. In this case, a pressure applied on the soft material would reduce the cross-sectional area of the channel and then increase the resistance. The geometry of channel(s)—for example, a serpentine arrangement—may increase the overall working range.

4.5.2   Electro-Active Polymers as Soft Robot Sensors

The EAP in section 4.3.3 can actually work as a sensor, using the very same capacitive principle in a reverse way. When a pressure is applied, the two electrodes get closer to each other and the voltage increases. From equation (4.23) and equation (4.24), you can extract the voltage as a function of the force applied and the stress, respectively.

4.5.3   Magnetic Phenomena for Soft Sensors

Similarly to the use of additives described in section 4.5.1, a dispersion of magnetized microparticles can be added to soft materials. They create a local magnetic field that can be detected by a magnetometer placed in close proximity to it. When a pressure is applied and the material deforms, the particles inside it move and rotate, and the magnetic field changes. Measuring this change gives a measure of the pressure applied.

4.5.4   Optical Principles for Soft Sensors

The way optical principles can be used for soft sensing is by measuring the changes of light intensity as it travels inside a soft material. Basic ingredients are a light source, a modulator, a transmitter, and a photosensitive detector such as a camera or a photodiode. Waveguides made of a light-transmitting material are inserted into the soft material in such a way that the light is fully reflected internally and detected on the other side of the waveguide. When the material is deformed on an applied pressure, the refractive index changes, and the optical signal is partially lost. The light intensity measured on the other side gives a measure of the deformation and the applied pressure.

4.5.5   Electrical Impedance Tomography (EIT)

This technique uses a sheet of a conductive material with electrodes placed on its boundaries. An electrical current is injected in the sheet, and the impedance distribution is observed through the electrodes. When a pressure is applied on the surface, the impedance distribution changes, and a tactile image is formed. Processing the signals from each pair of numerous electrodes is computationally intensive, but the tactile information obtained is rich, and sensors can be built in any shape.

4.6   Deformable Structures

In chapter 1, we defined soft robotics in two parts: robots built with either soft materials or deformable structures. The scope of this chapter, and this book, is the first part of our definition, but we mention a few basic approaches related to the second part. Specific cases of deformable structures are tensegrity structures, metamaterials, origami, and kirigami robots.

Tensegrity structures are a combination of stiff struts and a network of flexible tendons. Normally, tensional prestress ensures that tendons are always in tension, to maintain structural integrity. At the same time, the structure has the flexibility to deform under external interaction forces, still maintaining mechanical stability when the stress on the structure increases. As tendon tension increases, the structure becomes stiffer. For our soft robotics purposes, actuators are added to tune the tendon tensions and to change the strut angles. This way, deformations are obtained by internal interaction forces and used for robot movements, like locomotion. Study elements are the structure modeling and programmability, the actuation technologies to embed in the structure, and the materials for struts and tendons, which are sometimes smart materials with actuation functions.

Metamaterials are structures made from assemblies of multiple elements, and their mechanical properties are defined by their structure rather than their composition. Among their unusual properties is a negative Poisson's ratio, meaning that the increase of one dimension generates an increase in the other directions instead of a decrease.

Origami is the Japanese art of folding paper, and *kirigami* is a variation in which the paper is cut as well as folded. They can help soft robotics by suggesting morphing methods. Generally speaking, a flat sheet with embedded electronics can turn into one or more 3D morphologies to become a robot and move. Several well-known origami forms can be used in soft robotics, with diverse technologies for actuating the creases, ranging from SMA to pneumatic actuators. Modeling and programmability are key factors, as well as foldable materials and components.

In addition to the cases above, any design of structures that are deformed by external or internal interactions are good candidates for soft robots; refer again to figure 1.1 (b).

4.7   Summary

After recalling the main material mechanical properties, we have identified the soft ones as those with a Young's modulus below 10^9 Pa, thus identifying rubber, silicones, and hydrogels. We then walked through the actuation technologies for soft robotics to understand the working principles and learn the basic mathematical relations that describe them. We talked about stiffening mechanisms, which can be coupled with actuation, like the actuator antagonist arrangement, or can be completely independent components, like the jamming chambers added to soft robots. We also went through the sensing technologies for soft robotics, focusing on the working principles. Lastly, for consistency with our soft robot definition, we outlined a few possible approaches to deformable structures.

Self-Assessment Questions

1. In an EAP actuator, how can the output force be increased?

2. Propose a shape for an SMA actuator that is different from the spring shape.

3. Propose a concept for the use of layer jamming in a cylindrical soft arm.

4. How does one detect and measure a soft arm bending?

5. Propose your own design for a robot hand's soft fingers, which have to bend in order to grasp an object.

Further Readings

On Material Mechanical Behavior

Grote, Karl-Heinrich, and Hamis Hefazi, eds. 2021. *Springer Handbook of Mechanical Engineering*. 2nd Edition, Springer Nature Switzerland, Cham.

On Metamaterials and Origami Robots

Ahmad, Rafsanjani, Katia Bertoldi, and André R. Studart. 2019. "Programming Soft Robots with Flexible Mechanical Metamaterials." *Science Robotics* 4 (29): eaav7874. https://doi.org/10.1126/scirobotics.aav7874.

Jiao, Pengcheng, Jochen Mueller, Jordan R. Raney, Xiaoyu Zheng, and Amir H. Alavi. 2023. "Mechanical Metamaterials and Beyond." *Nature Communication* 14 (1): 6004. https://doi.org/10.1038/s41467-023-41679-8.

Rus, Daniela, and Cynthia Sung. 2018. "Spotlight on Origami Robots." *Science Robotics* 3 (15): eaat0938. https://doi.org/10.1126/scirobotics.aat0938.

On Soft Actuation Technologies

Leal, Pedro B. C., and Marcelo A. Savi. 2018. "Shape Memory Alloy-Based Mechanism for Aeronautical Application: Theory, Optimization and Experiment." *Aerospace Science and Technology* 76: 155–163. https://doi.org/10.1016/j.ast.2018.02.010.

Lee, Chiwon, Myungjoon Kim, Yoon Jae Kim, Nhayoung Hong, Seungwan Ryu, H. Jin Kim, and Sungwan Kim. 2017. "Soft Robot Review." *International Journal of Control, Automation and Systems* 15: 3–15. https://doi.org/10.1007/s12555-016-0462-3.

Pons, José L. 2005. *Emerging Actuator Technologies: A Micromechatronic Approach*. Chichester, UK: John Wiley & Sons. https://doi.org/10.1002/0470091991.

Zaidi, Shadab, Martina Maselli, Cecilia Laschi, and Matteo Cianchetti. 2021. "Actuation Technologies for Soft Robot Grippers and Manipulators: A Review." *Current Robotics Reports* 2: 355–369. https://doi.org/10.1007/s43154-021-00054-5.

On Soft Sensing Technologies

Roberts, Peter, Zadan Mason, and Carmel Majidi. 2021. "Soft Tactile Sensing Skins for Robotics." *Current Robot Reports* 2: 343–354. https://doi.org/10.1007/s43154-021-00065-2.

On Stiffening Technologies

Manti, Mariangela, Vito Cacucciolo, and Matteo Cianchetti. 2016. "Stiffening in Soft Robotics: A Review of the State of the Art." *IEEE Robotics and Automation Magazine* 23 (3): 93–106.

On Tensegrity Structures

Shah, Dylan S., Joran W. Booth, Robert L. Baines, Kun Wang, Massimo Vespignani, Kostas Bekris, and Rebecca Kramer-Bottiglio. 2022. "Tensegrity Robotics." *Soft Robotics* 9 (4): 639–656. https://doi.org/10.1089/soro.2020.0170.

5 Soft Robot Modeling

Chapter Objectives

· To understand the purpose of modeling soft robots

· To learn the techniques for modeling soft robots and their deformations as generated by actuators

· To learn the techniques for modeling the deformations of soft robots given by external interactions

5.1 Overview

Modeling a phenomenon or a system means giving a mathematical description of it. This is usually accomplished through equations that formalize the relations among the parameters that come into play. In robotics, modeling has generally two main purposes: control and design. In chapter 2, you see the main models used in robotics for control purposes. They are based on rigid-body physics, like kinematics, differential kinematics, and dynamics. In soft robotics, we need to release the rigid-body physics assumption, and therefore we cannot use the same modeling techniques. In addition, in chapter 1, we see that interactions with the environment are crucial to building intelligent behavior. This embodied intelligence paradigm is relevant for control because we include external interactions in the control system, reducing the overall computational burden. At the same time, a mathematical description of the external interactions provides indications for designing robots that take advantage of embodied intelligence.

For those reasons, in this chapter we are going to address both the soft robot movements stemming from actuation and the soft robot deformations given by external interactions.

5.2 One Equation for Modeling Soft Robots

Before entering the maze of modeling techniques, let us anchor ourselves to some terminology and formalism to share along this chapter. As introduced in the previous section, we are going to learn the following:

1) the techniques for modeling the deformations of a soft robot under the forces generated by the action of its own actuators—termed *internal interactions*

2) the techniques for modeling the deformation of a soft robot under the forces generated by the interaction with the surrounding environment, either a solid or a fluid—termed *external interactions.*

We may represent them all inside one equation, according to the formalism adopted in (Mengaldo et al. 2022):

$$\mathcal{D}\mathbf{q}_{sb} = \mathcal{N}_{sb} + \mathcal{C}_{int} + \mathcal{C}_{ext} \quad \text{on } \Omega_{sb} \tag{5.1}$$

where $\mathcal{D}$ is a differential operator that can assume the form of a partial derivative ∂ or a total derivative d, of first or second order; $\mathbf{q}_{sb}$ describes the soft body (sb); $\mathcal{N}_{sb}$ is a nonlinear term describing the soft-body mechanics; and $\mathcal{C}_{int}$ and $\mathcal{C}_{ext}$ are two coupling terms accounting for internal and external interactions, respectively—all on the soft-body domain Ω_{sb}.

For the moment, let us take the general meaning of the equation (5.1) terms. We are going to see how each of them is instantiated, depending on the modeling approach adopted.

5.3 Modeling Internal Interactions

As always (see also chapter 2), modeling a robot is a two-sided coin. We can find a *direct* or *forward model* that computes the robot position from motion or forces in the actuator space, or we can find an *inverse model* that computes what actuators should do in order to achieve a desired robot position. We are going to focus on the direct modeling case for the rest of this section.

Modeling internal interactions means modeling the soft-body deformations, modeling the physics of actuation, and coupling them. As you see, we are referring to the first two terms of the right side of equation (5.1), namely $\mathcal{N}_{sb}$ and $\mathcal{C}_{int}$. We are going to define the forms they assume.

As a general classification of the techniques described in the following subsections, we may consider the two broad realms of 3D continuum solid mechanics and finite-dimensional spaces. We will see in the next section that such a broad classification applies to external interactions as well. In the first case of continuum mechanics, the mathematical solution consists of solving sets of partial differential equations (PDEs). In the second case, we will need to solve set of ordinary differential equations (ODEs) of multibody dynamics.

5.3.1 Three-Dimensional (3D) Continuum Solid Mechanics Models

The deformation of a soft robot can be described in the realm of 3D continuum mechanics. Among well-known tools are finite element methods (FEMs). In this case, the geometry of the robot shape is described by a mesh that consists of a set of nodes and connections between neighboring nodes, as illustrated in figure 5.1. The deformations are reflected in the change of the mesh geometry. The volume can be interpolated from the node positions.

A full FEM course is beyond the scope of this book. FEM is a common tool in engineering, and the take-home message here is that using this very same tool for modeling soft robots and their deformations is in fact possible. FEM approaches are especially suitable to take into account volume variations. Recalling the diverse types of actuation technologies described in chapter 4, you may quickly tell that such an approach is well suited in the case of fluidic actuation.

What happens to our equation (5.1) terms when we apply this 3D continuum mechanics approach? First of all, let us remove the C_{ext} term for the moment, accounting for the external interactions. $\mathcal{D}$ assumes the form of a partial derivative ∂ here, with respect to time t. $\mathbf{q}_{sb}$ becomes v_{sb}, which is the velocity field of the soft body (sb). $\mathcal{N}_{sb}$ describes the soft-body mechanics as the divergence of the soft-body stress tensor, $\nabla \cdot \sigma_{sb}$, as follows:

$$\frac{\partial}{\partial t} v_{sb} = \nabla \cdot \sigma_{sb} + C_{int} \tag{5.2}$$

How can we introduce the actuation effect represented by C_{int}? Actuation is imposed here as equality and inequality constraints, $\mathfrak{T}_e$ and $\mathfrak{T}_i$, through available tools like Lagrange multipliers $C_{int} = \Lambda^T \mathfrak{T}_e + \Theta^T \mathfrak{T}_i$. Equation (5.1) then becomes:

$$\frac{\partial}{\partial t} v_{sb} = \nabla \cdot \sigma_{sb} + \Lambda^T \mathfrak{T}_e + \Theta^T \mathfrak{T}_i \tag{5.3}$$

5.3.2 Rod Models

When our soft robot has an elongated morphology, with two dimensions much smaller than the third one, similar to the one depicted in figure 1.1(a), rod models can be used. In this case, we can consider the soft robot as a rod, thin and continuous. At this point, describing our soft arm deformation is as simple as describing the position and orientation of the points of its backbone. At any point along the backbone, and at any time instant, such deformation is described by a function that represents the position and orientation of a reference frame centered in that point. Figure 5.3 may help you to grasp this general concept. Conversely from the FEM approach, rod models are suited when volume variations are negligible. Recalling chapter 4, you may think of cable actuation, for instance.

Let $\mathcal{F}(s)$ be the function describing the spatial curve of our soft robot backbone. s is in the space of normalized rod length [0, 1], where 0 means the rod base and 1 the rod tip, as indicated in figure 5.3. The rod deformations can be described by local strains that represent the variations of $\mathcal{F}$ at any infinitesimal point $s + \varepsilon$, with respect to a previous point s. Figure 5.2 illustrates a few cases, such as elongation, bending, torsion, and shear

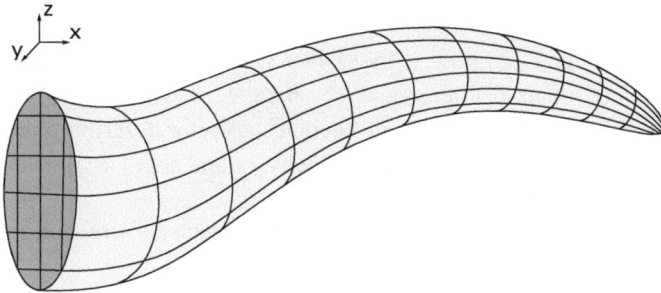

Figure 5.1
A mesh describes the geometry of the rod-like robot through a set of nodes and their connections.

Not deformed

Elongation

Torsion

Shear strain

Curvature

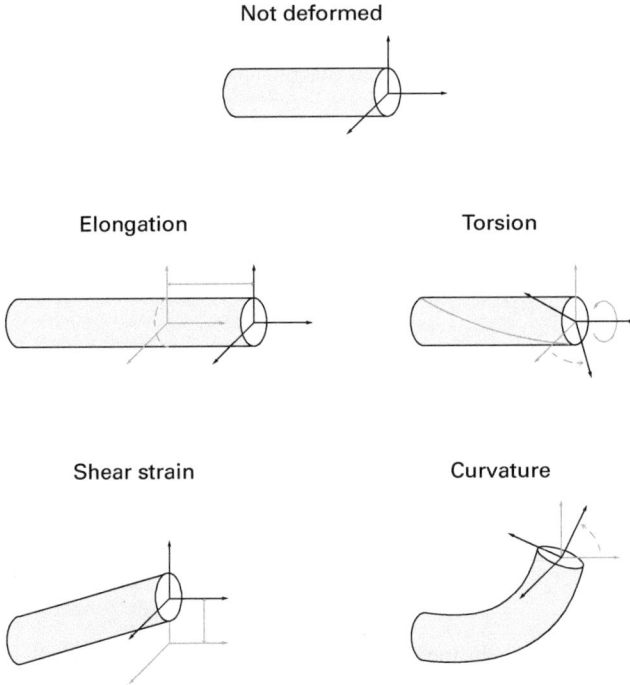

Figure 5.2
Rod deformations described as local strains. The variations of the position and orientation of the reference frame are illustrated with respect to a previous point at infinitesimal distance.

strain. You may recognize here a sort of forward kinematics, from chapter 2, but in the continuum domain.

A general modeling approach that can be used with rods is the Cosserat's model. In this model, a rod is a continuous set of infinitesimal micro-solids stacked along the backbone. A Cosserat's rod is then modeled as a continuous set of cross-sectional frames centered on each micro-solid, which can be formalized as follows:

$$\mathcal{F}(s) = (O, t_1, t_2, t_3)(s) \tag{5.4}$$

where $O(s)$ is on the backbone, $t_1(s)$ is a unit normal vector perpendicular to the cross-section, and $t_2(s)$ and $t_3(s)$ are the unit vectors spanning the cross-sectional plane, with $s \in [0, 1]$.

Let us go back to equation (5.1) and see how it looks like when we use Cosserat's models. Again, we do not consider external interactions C_{ext} here. $\mathcal{D}$ assumes again the form of a partial derivative ∂, with respect to time t. $\mathbf{q}_{sb}$ becomes a vector describing the velocity twist field of the soft body sb,[1] containing linear velocity $\mathbf{v}_{sb}$ and angular velocity ω_{sb}: $\begin{bmatrix} v_{sb} \\ \boldsymbol{\omega}_{sb} \end{bmatrix}$. $\mathcal{N}_{sb}$ becomes a vector as well describing the soft-body mechanics related to linear and angular velocities: $\begin{bmatrix} \mathcal{N}_{sb,v} \\ \mathcal{N}_{sb,\omega} \end{bmatrix}$.

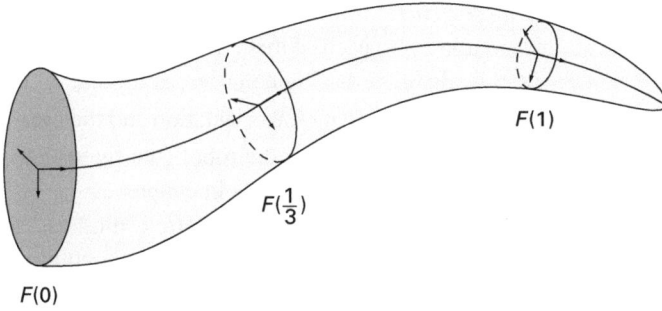

Figure 5.3
In rod models, deformations are described as the positions and orientations of the points of the backbone—that is, the central axis. This means that a given value of $\mathcal{F}$ describes a point at a distance sL from the base, given L as the total rod length. In the Cosserat's model, the rod is a continuous set of infinitesimal micro-solids stacked along the backbone and modeled as a continuous set of cross-sectional frames centered on each micro-solid, as expressed in equation (5.4). The cross-sections are considered undeformable.

$$\frac{\partial}{\partial t}\begin{bmatrix} v_{sb} \\ \omega_{sb} \end{bmatrix} = \begin{bmatrix} \mathcal{N}_{sb,v} \\ \mathcal{N}_{sb,\omega} \end{bmatrix} + \mathcal{C}_{int} \tag{5.5}$$

These $\mathcal{N}$ terms depend on a number of parameters, such as the density ρ of the material, the momentum of inertia I, the area of the cross-section A, the position of the frame $O(s)$, and the linear and angular cross-sectional stresses along the rod. Actuation $\mathcal{C}_{int}$ is introduced in this case as active internal wrenches.[2]

5.3.3   Finite Parametrization Models

So far, despite learning about quite different modeling approaches, we have kept ourselves inside the realm of continuum mechanics. As mentioned in section 5.3, we must solve PDEs in this realm. Instead, moving outside, toward spaces of finite-dimensional parameters, would bring us to more conveniently solve ODEs. This big move implies giving up modeling the full details of our soft robot deformations and instead making approximations that can still give us an accurate-enough description.

You may think it is a bit drastic, but in some cases, still considering our rod-like soft robot of figure 1.1(a), we can just assume a constant-curvature (CC) for the whole arm. This is correct when the manipulator is uniform in shape and symmetric in actuation design, when external loading effects are negligible, and when torsional effects are negligible. Under this modeling approach, the robot configuration is fully described by one curvature and its direction.

We may take it a step further and consider a piecewise constant-curvature model (PCC). In soft robots that are built with two or more modules, we can apply the CC modeling approach to each module and connect them together. We assume that the curvature is piecewise constant and that discontinuities only appear at some fixed nodes. Two subsequent nodes are then connected with an arc of circumference described by the curvature of the module between them. All deformations are neglected but bending. This sequence of curvatures can be seen as the soft robot's degree of freedom (DOF).

However, those are all special cases of a more general formulation that sets a discretional finite number of points along the backbone and assigns to each of them a frame $\mathcal{F}$ describing a rigid-body displacement. Each frame origin is along the backbone curve, at a point s_i. $\mathcal{F}_{s_i}$ is the transformation (rotation and translation) matrix describing the orientation and the translation of the frame with respect to the world frame at the base of the robot (see figure 5.4). Again, you may find some similarities here with the kinematic models in chapter 2.

Going back to equation (5.1), with finite-dimensional parametrization models, $\mathcal{D}$ assumes the form of a total derivative d, with respect to time t. $\mathbf{q}_{sb}$ is a vector containing the curvature of the soft body sb, with respect to arc length s. $\mathcal{N}_{sb}$ becomes a function of the soft-body displacement $\mathbf{s}_{sb}$. The actuator forces contained in $\mathcal{C}_{int}$ are grouped in a vector α that acts as a forcing term. Equation (5.1) then becomes:

$$\frac{d}{dt}\mathbf{q}_{sb}(s_i) = -\mathbf{M}^{-1}(\mathbf{D}(\mathbf{s}_{sb}, \mathbf{q}_{sb}) + \mathbf{K}(s_{sb})) + \alpha \qquad (5.6)$$

The nonlinear term just introduced is derived from the robot dynamic model (see also chapter 2), where $\mathbf{M}$ is the inertial matrix, $\mathbf{D}$ is a dissipative term that includes internal friction and other forces (such as Coriolis forces), and $\mathbf{K}$ is an elastic term that encapsulates the stiffness of the system.

5.4   Modeling External Interactions

It is now time to put the $\mathcal{C}_{ext}$ term back in the equation—that is, we want to see how we can model the contribution of external interaction forces to the deformations of our soft robot.

First of all, it is useful to clarify what we mean by external interactions and the kind of external surroundings we are considering for our soft robots and why they are relevant for our robot behavior. Rather than attempting to consider all possible soft robot morphologies, abilities, and operational environments, we want to instead stick to using figure 1.1 as a reference.

• **Solid interactions**. It is quite straightforward to consider our soft robot as operating in the air and interacting with solid materials, like the objects it grasps or the terrain it moves on

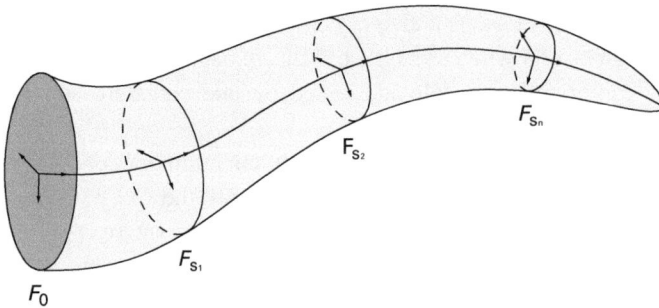

Figure 5.4
In finite parametrization models, a reference frame is set on a finite number of points along the backbone. Transformation (rotation and translation) matrices describe the orientation and the translation of each frame with respect to the base frame.

(see figure 5.5). In manipulation and locomotion tasks, interactions with solids prevail. The forces involved in interacting with the object play a crucial role in grasping in a bidirectional way: the forces on the soft robot gripper help it deform and adapt to the object's shape, thus achieving a more effective configuration; the gripper forces on the object enable stable grasping and lifting. In locomotion, what produces a thrust forward is the interaction with the solid substrate and the friction associated with the interaction forces.

· **Fluid interactions**. Another relevant case is the use of soft robots underwater, where they interact with a liquid and similarly with solid objects and substrate. Here, the interactions with the fluid become prevalent (see figure 5.6). Movements in water are affected by the forces exchanged with the fluid, whether helpful to the task or unwanted. For example, when swimming, we take advantage of the thrust produced by the interaction with water, which propels us forward. By contrast, moving an arm or a leg underwater creates water resistance. Walking underwater involves a combination of fluid interactions and interactions with the solid terrain.

For completeness, granular materials would complement the picture, and some soft robots can actually burrow and move underground. However, this section's scope is limited to interactions with solids and fluids, which encompasses the vast majority of soft robots in existence today.

Similar to the case of internal interactions, we may devise techniques in the realm of continuum mechanics and techniques based on finite, lumped parameters. The following sections explain the main conceptual approaches, pointing readers to more extensive resources on related math and techniques that are general and common to other engineering fields and not specifically adapted to the case of soft robotics.

5.4.1 Continuum Fluid Mechanics Models

Navier-Stokes equations are the main tool for describing flow physics in the continuum domain. They describe the behavior of fluid particles based on conservation of mass, momentum, and energy. We are interested in the action of the fluid on our soft robot, thus we are interested in what happens at the interface between the fluid and soft body $\Gamma_{sb,f}$. There, we can identify equality constraints, as follows:

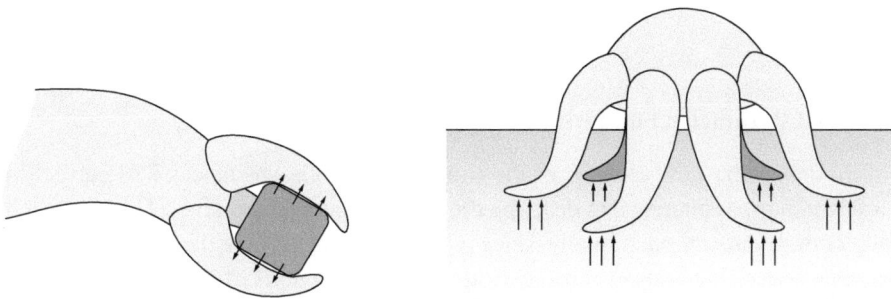

Figure 5.5
Interaction with solids is relevant for grasping and for locomotion. Reaction forces from objects and friction are important for grasping objects. Locomotion is based on ground reaction forces and friction as well.

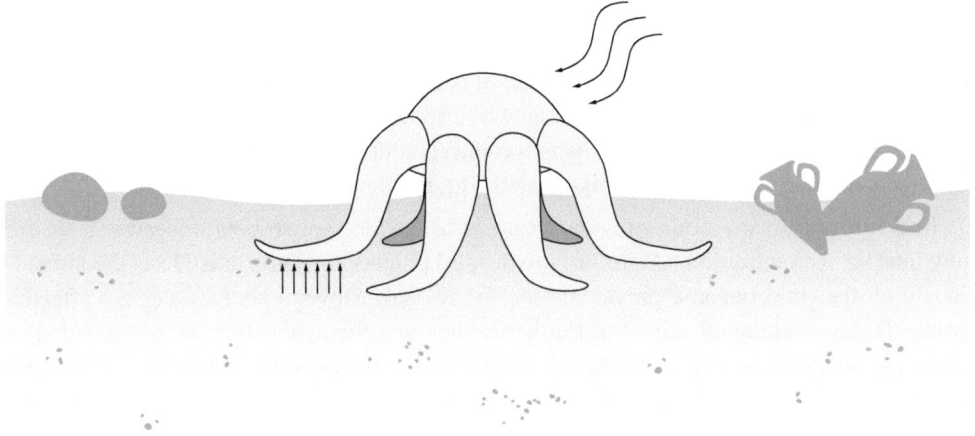

Figure 5.6
For soft robots operating in water, interaction with the surrounding fluid affects both arm movements and locomotion. Water drag applies forces on the soft arm, contributing to its deformation during arm motion in water. Buoyancy affects locomotion in water, which is generally obtained with longer flight phases for this reason.

$$\mathfrak{T}_{e,f} = \begin{cases} \mathbf{q}_{sb} = \mathbf{q}_f \\ \sigma_{sb} \cdot \mathbf{n} = \sigma_f \cdot \mathbf{n} \\ \mathbf{x}_{sb} = \mathbf{x}_f \end{cases} \tag{5.7}$$

In the first equality constraint, $\mathbf{q}_{sb} = \mathbf{\nu}_{sb}$ corresponds to the velocity of the soft body and $\mathbf{q}_f = \mathbf{\nu}_f$ is the velocity of the fluid at the interface $\Gamma_{sb,f}$. The velocities need to be equal at the interface, as we wish the relative velocity to be zero at the interface. In the second equality constraint, $\mathbf{n}$ is the normal at any soft-body location, and σ_f is the fluid Cauchy stress tensor composed of the sum of the deviatoric stress tensor $\tau^{(f)}$ that accounts for the viscosity and a pressure term $-p\mathbf{I}$ ($\mathbf{I}$ is the identity matrix). We need the normal stress tensors to be equal at the interface. Finally, in the third constraint, we have equal positions of the soft body $\mathbf{x}_{sb}$ and the fluid $\mathbf{x}_f$ at the interface.

Such equality constraints can be inserted in equation (5.1) by expressing the $\mathcal{C}_{ext}$ term[3] as Lagrange multipliers $\Lambda^T \mathfrak{T}_{e,f}$:

$$\mathcal{C}_{ext} = \Lambda^T \mathfrak{T}_{e,f} \tag{5.8}$$

5.4.2 Lumped Parameter Fluid Models

Instead of modeling the full behavior of the fluid, in some circumstances it is correct to identify a few major parameters that describe the behavior at a macro level. This approach is generally convenient computationally, since we do not need to solve the fluid mechanics equations seen above. We instead represent our system with a set of discrete parameters connected or lumped together. In the case of fluids, relevant parameters are *added mass*, *drag*, *lift*, and *buoyancy*.

When a solid object moves inside a fluid, the fluid exerts a force on the solid because of its inertia, which is proportional to the mass of the fluid that is displaced by the solid.

This force is known as the added mass force, and we name it $\mathbf{f}_{\text{added mass}}$. The added mass is a virtual mass or inertia added to the soft robot body moving in the surrounding fluid. For instance, if our robot would be perfectly spherical and immersed in an incompressible fluid, the added mass would be represented as follows:

$$\mathbf{f}_{\text{added mass}} = \frac{1}{2}\rho_f V_{\text{sb}}\left[\frac{D\mathbf{q}_f}{Dt} - \frac{d\mathbf{q}_{\text{sb}}}{dt}\right] \tag{5.9}$$

where $\dfrac{D\mathbf{q}_f}{Dt}$ is the material derivative of the fluid velocity, $\dfrac{d\mathbf{q}_{\text{sb}}}{dt}$ is the total derivative of the velocity of the spherical soft body, ρ_f is the fluid density, and V_{sb} is the volume of the spherical soft body.

While moving inside the fluid, the object also creates a disturbance in the fluid flow. This disturbance results in a pressure difference between the front and back of the object. The pressure at the front of the object is higher than the pressure at the back of the object, which creates a net force that opposes the motion of the object. This force is called drag, and we name it $\mathbf{f}_{\text{drag}}$. It is proportional to the velocity of the fluid flowing around the soft body and is expressed in the following equation:

$$\mathbf{f}_{\text{drag}} = \frac{1}{2}\rho_f \mathbf{q}_f A_{\text{sb}} C_{\text{drag}} \tag{5.10}$$

where A_{sb} is the area of the soft body exposed to the fluid and C_{drag} is a drag coefficient, which is typically tabulated for simple geometries.

Another force that is created when a solid object moves inside a fluid is lift, which is perpendicular to the direction of motion and proportional to the velocity of the fluid flowing around the soft body. We name it $\mathbf{f}_{\text{lift}}$ and express it in a similar way, by considering a C_{lift} coefficient:

$$\mathbf{f}_{\text{lift}} = \frac{1}{2}\rho_f \mathbf{q}_f A_{\text{sb}} C_{\text{lift}} \tag{5.11}$$

Finally, buoyancy is the net vertical force caused by the difference in pressure between the top and bottom of the object. It is proportional to the volume of the fluid displaced by the object. We use $\mathbf{f}_{\text{buoyancy}}$ here, also accounting for gravity.

Once we have identified the lumped parameters that fully describe our soft robot in a fluid, we can include the external interaction forces $\mathcal{C}_{\text{ext}}$ in equation (5.1) by simply summing up the difference force contributions:

$$\mathcal{C}_{\text{ext}} = \mathbf{f}_{\text{added mass}} + \mathbf{f}_{\text{drag}} + \mathbf{f}_{\text{lift}} + \mathbf{f}_{\text{buoyancy}} \tag{5.12}$$

5.4.3 Continuum Solid Mechanics Models

As mentioned at the beginning of section 5.4, interactions of our soft robot with external solids are relevant for both manipulation and locomotion tasks. We are interested in capturing the normal and tangential forces as well as the friction and adhesion forces that affect the soft body and can be leveraged for accomplishing the tasks.

In the realm of continuum mechanics, modeling external solid interactions involves the same techniques used for modeling internal interactions, as described in section 5.3.1. FEM can be used to describe a surrounding solid medium, in addition to describing the soft robot and its internal interactions.

Similar to the case of external interactions with fluids, the $\mathcal{C}_{ext}$ term of the equation consists of a set of constraints to satisfy at the interface between the soft body and the solid medium. In addition, we have to account for the transitions from no contact to contact or from static friction to dynamic friction.

5.4.4   Lumped Parameter Solid Models

As explained in section 5.4.2, proper lumped parameters can effectively serve the purpose of modeling the overall behavior of a system, avoiding the computational burden of accurate continuum modeling. In the case of solid external interactions, we may aggregate the effect of the forces shaping the contact between two surfaces.

First of all, the state $\mathbf{q}_s$ of a solid can be described by the position and orientation of its center of mass and its linear and angular velocities. We then need to identify the point of contact between the soft body and the rigid solid and then apply an equivalent frictional force $\mathbf{f}$ and torque τ, induced by the interaction between the soft and the rigid body. The interaction surface between the rigid solid and the soft robot is typically modeled as a planar surface. This choice leads to models that contain only the three DOFs associated to frictional forces at the planar surface. In practice, a soft robot interacting with a rigid body has nonplanar contact surfaces with multiple points of contact. As a result, there are three additional DOFs, and the model for the normal and frictional wrenches is 6D, as both force $\mathbf{f}$ and torque τ are 3D. We can express the external interaction term as such:

$$\mathcal{C}_{ext} = \mathbf{f}_n + \tau_n + \mathbf{f}_t + \tau_t \tag{5.13}$$

where $\mathbf{f}_n$ is normal force, τ_n is resulting torque, $\mathbf{f}_t$ is frictional force and τ_t is resulting torque.

5.5   Reduced-Order Models

As you understand intuitively, the more precise our robot mathematical description is, the more complex the computational task of modeling and simulating our robot becomes. Using soft robot models for designing and for controlling them is a balance between computational complexity and completeness of models. Sometimes, especially for modeling the macroscopic interactions of our robot with the external environment, we may wish to focus on some fundamental relations. We may wish to reduce the model to the relations between some major parameters describing the soft robot and some major parameters describing the external environment. Such reduced-order models (ROMs) do not intend to describe the behavior of every single, detailed part of the robot body; rather, they approximate it with masses concentrated in one point, punctual forces, and spring-damper systems, for example.

Let us move to locomotion and consider a soft robot leg contacting the ground. Our model here can approximate the robot body just with its mass, located in its center of gravity, and each leg as a spring-damper system, receiving reaction forces from the ground. These are also called *fundamental models*, and they help describe and design the robot behavior, including its interaction with the environment and hence its embodied intelligence.

5.6 Data-Driven Approaches

It is worth mentioning that data-driven approaches are used for solving the same modeling problems that we have seen in the previous sections. As an example, neural networks are used to learn the mapping of the previous to the next state directly, thus transforming equation (5.1) into this expression:

$$\mathcal{D}\mathbf{q}_{sb} = NN[\mathbf{q}_{sb}(k) \rightarrow \mathbf{q}_{sb}(k+1)] \tag{5.14}$$

where k are different time instances.

The techniques used for implementing such approaches are out of the scope of this chapter.

5.7 Summary

We know that modeling is necessary in robotics, both for control and for design purposes. In robot control, we especially need models of the action of actuators on the robot movements. This chapter has explained that soft robot models are similarly necessary both for control and for design, and we similarly need models of internal interactions (i.e., models of the effect of actuators on robot deformation). In order to introduce embodied intelligence when designing soft robots, we also need models of external interactions (i.e., of the external forces acting on the robot and deforming its soft, compliant body).

We anchored ourselves to a rod-like soft robotic arm to explain the main techniques that can be used for modeling internal and external interactions, either in the realm of continuum mechanics and in discrete spaces of relevant parameters. Such theory and techniques can be generalized to other possible soft robot morphologies.

We mentioned that a trade-off is sometimes needed between an accurate mathematical description and an affordable computational burden that may otherwise hinder a practical use of models. The strategy of building ROMs from physics-related major parameters responds to this need and provides helpful models for robot functions like locomotion.

Finally, we mentioned how the same modeling problems can be solved with data-driven approaches, although we stop short of venturing into that realm.

Self-Assessment Questions

1. Which technique would you use for modeling the internal interactions of your silicone rod-like robotic arm actuated pneumatically?

2. Same as question 1, which technique would you use if your arm has a tendon-driven actuation?

3. Your robot is inspired by an octopus and has a main soft body with eight arms that are used both for grasping and for locomotion. Which of the modeling techniques seen before are relevant to your case?

4. Can you apply a rod model to a soft robot leg?

5. What parameters would you consider relevant for underwater locomotion?

Further Readings

On Basics of Solid Mechanics

Lubliner, Jacob, and Panayotis Papadopoulos. 2014. *Introduction to Solid Mechanics: An Integrated Approach.* Springer.

On Data-Driven Methods and Reduced-Order Models

Brunton, Steven L., and J. Nathan Kutz. 2022. *Data-Driven Science and Engineering: Machine Learning, Dynamical Systems, and Control*, 2nd ed. Cambridge: Cambridge University Press. Chapter 12 in this edition is recommended for reduced-order models.

On Rod Models

Armanini, Costanza, Frédéric Boyer, Anup Teejo Mathew, Christian Duriez, and Federico Renda. 2023. "Soft Robots Modeling: A Structured Overview." *IEEE Transactions on Robotics* 39 (3): 1728–1748. https:/doi.org/10 .1109/TRO.2022.3231360.

On Soft Robot Simulators

Coevoet, Eulalie, Thor Morales Bieze, Frederick Largilliere, Zhongkai Zhang, Maxime Thieffry, Mario Sanz Lopez, Bruno Carrez, Damien Marchal, Olivier Goury, Jeremie Dequidt, and Christian Duriez. 2017. "Software Toolkit for Modeling, Simulation, and Control of Soft Robots." *Advanced Robotics* 31 (22): 1208–1224. https:// doi.org/10.1080/01691864.2017.1395362.

Mathew, Anup Teejo, Ikhlas Ben Hmida, Costanza Armanini, Frédéric Boyer, and Federico Renda. 2023. "SoRoSim: A MATLAB Toolbox for Hybrid Rigid—Soft Robots Based on the Geometric Variable-Strain Approach." *IEEE Robotics and Automation Magazine* 30 (3): 106–122. https://doi.org/10.1109/MRA.2022 .3202488.

6 Soft Robot Control

Chapter Objectives

- To understand the control problems in soft robotics
- To translate the concepts of robot control recalled in chapter 2 into soft robot control
- To approach learning-based control methods

6.1 Overview

Controlling soft robots requires some adaptation of the techniques recalled in chapter 2. There are a few dramatic differences in the fundamental descriptions of a soft robot position and in the way space transformations are done. This chapter sets the key concepts for transitioning to techniques that comply with soft robot characteristics. Then we can conceptualize control techniques for soft robots, in joint and task space, making use of the models learned in chapter 5. We also explore alternative approaches that do not make use of models but employ learning techniques for building the required transformations and control relations with so-called neurocontrollers.

To circumvent the variety of morphologies of soft robots, we remain anchored to the soft robot arm of figure 1.1(a). It is a simple planar case with one actuator and is helpful for grasping the concepts that can be generalized to more complex cases.

6.2 Soft Robot Control Problems

In chapter 2, we recalled the expression of a robot position in Cartesian space as a vector $\mathbf{x}$ containing a position and an orientation, each expressed with three parameters. This holds for a soft robot as well—for instance, for the position of the tip of our soft robot arm in figure 1.1(a). In a rigid-link robot with revolute joints, such a position normally corresponds to a position in joint space, expressed by a vector $\mathbf{q}$ containing the value of each joint angle. This space normally overlaps with the actuator space in a traditional configuration with a motor controlling each joint. Here, it has already become difficult to find a good match with our soft robot. The position $\mathbf{x}$ is given by the curvature of the arm, as generated by the internal forces that the actuators produce. The curvature is the robot configuration that we call $\mathbf{k}$ hereafter. We have to consider the actuator space separately, and we can call it $\mathbf{q}$ as in the case we know.

The relations and the transformations between those spaces are key for robot control. In chapter 2, we have the kinematics relations that transforms $\mathbf{q}$ in $\mathbf{x}$ (forward kinematics, equation (2.3)) and vice versa (inverse kinematics, equation (2.4)). We need a few more transformations here, from $\mathbf{q}$ to $\mathbf{k}$ and then to $\mathbf{x}$, and the inverse way back, if possible. It is worth noticing that, as in the case of kinematics transformations of chapter 2, inverse transformations are not always possible, or there may be infinite solutions for bringing the end effector to the same position. In soft robots, this is even more true. Soft robots are generally underactuated (i.e., the number of control signals is smaller than the number of degrees of freedom). However, for the sake of simplicity, we stick on our simple arm example here, and we assume that inverse transformations are solvable.

In chapter 4, we have different kinds of actuators generating a bending to our soft robot arm, as illustrated in figures 4.6, 4.7, 4.11, and 4.13. The inputs to the system are different, too: a motor torque that increases a cable tension T, in the cable-driven arm; a pressure P, in the pneumatically actuated arm; a current that heats a shape-memory alloy (SMA) spring that generates a force F, in the SMA-actuated arm; and a voltage that makes an electro-active polymer (EAP) stack contract and generate a force F, in the EAP-actuated arm. In all cases, those inputs generate a curvature in the arm. The relations between them and the curvature that they generate are specific for the arm characteristics. Chapter 5 describes the techniques for modeling such internal interactions that deform the soft body due to the actuator actions. We have the transformations from $\mathbf{q}$ to $\mathbf{k}$, then. The transformations from $\mathbf{k}$ to $\mathbf{x}$ are geometrical transformations from the arm curvature to the tip position in space. They are independent from the robot characteristics. Figure 6.1 summarizes the transformations described above, with specific reference to the actuation cases of chapter 4 as examples.

6.3   Soft Robot Controllers

In chapter 2, we recalled that, given a desired position $\mathbf{x}_d$, a controller can nullify the error with the current position, either in the joint space or in task space. Figures 2.10 and 2.12 illustrate robot control in joint space and task space, respectively. Let us try to adapt those schemes to our soft robot arm, once we have outlined the transformations involved.

6.3.1   Joint Space Control

In the general control scheme in joint space of chapter 2, the desired position $\mathbf{x}_d$ is transformed into a desired position $\mathbf{q}_d$ already outside the control loop so that the comparison with the current position can be done with $\mathbf{q}$. In our case, we do the same transformation by going first from $\mathbf{x}_d$ to $\mathbf{k}_d$ and then from $\mathbf{k}_d$ to $\mathbf{q}_d$. Both transformations are inverse with respect to what we have seen so far. The first one is a geometric transformation, independent of the arm characteristics, while the second transformation is the inverse of one of the models learned in chapter 5. Figure 6.2 shows the new control scheme. Let us leave the big CONTROL block to handle the error $\mathbf{q}_e$ and produce a control signal $\mathbf{u}$. The techniques by which it can be

Control signal	Actuator input	Actuator space	Configuration space	Task space
u	-	q	k	x

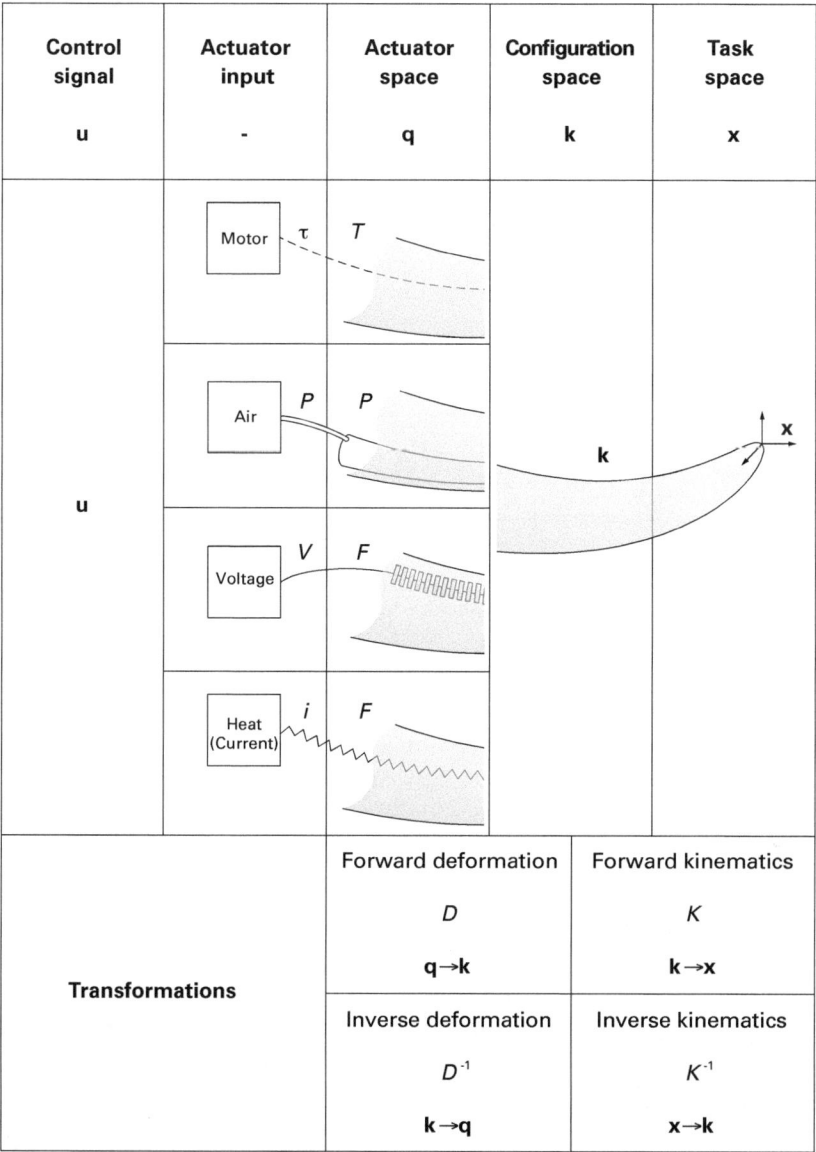

	Forward deformation	Forward kinematics
	D	K
	$q \rightarrow k$	$k \rightarrow x$
Transformations	Inverse deformation	Inverse kinematics
	D^{-1}	K^{-1}
	$k \rightarrow q$	$x \rightarrow k$

Figure 6.1
Transformations involved in the control of soft robots.

implemented are not different from those already used in robotics. We need to assume that we have sensors for the current **q**, such as cable tension sensors, pressure sensors, and force sensors, for instance. They give us the feedback for calculating the error $\mathbf{q}_e$.

6.3.2   Task Space Control

In task space control, as shown in figure 6.3, we close the control loop directly on the **x** positions, calculating the error $\mathbf{x}_e$ between the desired position $\mathbf{x}_d$ and the current **x**. There is no need to add anything outside the control loop now, but we need inverse transformations

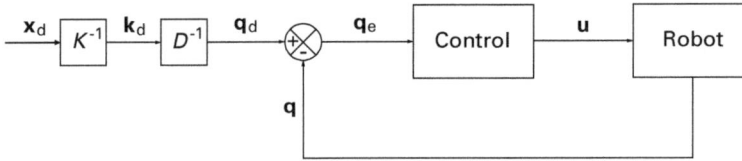

Figure 6.2
Joint space control for soft robot arms (adaptation of figure 2.10). Two transformations are needed outside the control loop. The ROBOT block includes the actuators that produce the **q** to the robot, as well as the sensors providing feedback on the current position.

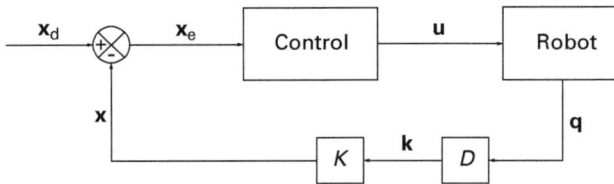

Figure 6.3
Task space control for soft robot arms (adaptation of figure 2.12). The CONTROL box contains inverse transformations from **x** to **k** and from **k** to **q**, while the feedback is transformed in a forward way from **q** to **k** and then to **x**.

inside the CONTROL box. Furthermore, the feedback for closing the loop is a **q**, and we need forward transformations to obtain an **x**.

As in the robot task space control of chapter 2, what is inside the CONTROL block is much more complex than the joint space case.

6.4   Soft Robot Neurocontrollers

In the control systems of chapter 2 and this chapter, some forms of models are used, either in their forward or inverse formulation. In the case of soft robots, those models are generally more complex to formulate, as we explain in chapter 5. Their accuracy may be limited by the inaccuracies of soft robots due to the materials and the fabrication techniques that are used. The computational complexity of the control system embedding those models tends to increase with the model complexity. In practical implementations, the controllers of section 6.3 may be limited to easier cases, like quasi-static conditions.

Alternative approaches are offered by *neurocontrollers*, which are controllers based on neural networks that use learning mechanisms to build up the relations that are needed to control a robot (see also section 3.8). For instance, to solve the classical control problem of producing a control signal **u** for the actuators, starting from a desired end effector position x_d, a neural network may replace the controller and give **u** as an output, after receiving x_d and the current **x** as an input. The network would learn how to produce the correct output during a training phase, in which the robot performs random movements and checks their effects in terms of end effector position.

Going a step backward and recalling section 3.8, a neural network is a computing technique inspired by the way neurons make computations in biological nervous systems. The same principle of connecting neurons in a network is used, with neurons producing a spike output when the weighted sum of its inputs reaches a threshold. The possibility of changing the weights and the connections is what makes learning possible.

Although many different typologies of artificial neural networks exist, common multi-layer networks have an input layer of neurons, an output layer, and a number of hidden layers in between them. The neurons in each layer can be connected to each neuron in the next layer. Recurrent neural networks also contain cycles. These recurrent connections provide a sort of memory to the network. Learning has three main paradigms: supervised, reinforcement, and unsupervised learning.

Supervised learning

As the name suggests, a supervisor is needed here because we need to know what output is expected for a given input. The error between the correct output and the actual one is used to correct the weights in such a way that the error is nullified or reduced. The weights of the output layer connections are corrected first, then the others are corrected sequentially until the input layer is reached. The error is propagated backward, and this learning technique is named backpropagation. The learning is repeated until convergence to a configuration in which no more changes are needed or a sufficiently low error is reached, smaller than a given threshold. After learning, the network gives correct outputs, generalizing to inputs that have not been seen during the training phase.

Reinforcement learning

In this case, the exact output value in each situation is not specified. Instead, the network has a way to know whether the output is good or not (i.e., it receives a positive reinforcement, a reward, or instead a negative reinforcement or a punishment, respectively). To give an example and explain the difference with supervised learning, let us consider a robot grasping an object. A reinforcement learning algorithm would explore the space of possible states; when the object is kept stably and does not drop, the robot receives a reward. With supervised learning, we should know the correct set of actions needed for a stable grasp.

Unsupervised learning

There is no external feedback to the network in this learning paradigm. The network creates a map of inputs and outputs autonomously, whichis why these kind of networks are said to have self-organizing maps. They are good at classification problems, where the network is presented with data to find trends. Mapping a robot task space into its joint space is also a task for self-organizing maps.

Going back to our control problem, we can replace a controller with a neural network as shown in figure 6.4.

6.5   Summary

This chapter makes us take the leap from robot control to soft robot control, outlining the aspects that are specific to soft robots and that make control techniques different. We have seen analogies and differences and set our spaces for the transformations on which control relies. Although we took a somewhat simplistic view of the adaptation of basic control

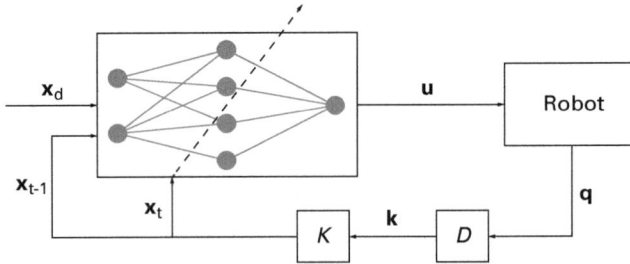

Figure 6.4
Illustration of a neurocontroller that solves the control problem of generating a control signal for a desired position, given the current status (position at previous time t-1). The feedback is used during training to associate the robot position reached after a movement with the corresponding movement control signal. The arrow on the network indicates learning. During training, the feedback can be given by direct transformations, as indicated here, or by purposive sensors such as motion tracking systems.

approaches to the soft robot case, it gives us a conceptual foundation on which to build the many soft robot control techniques that are increasingly available in the literature. From there, we realize the need for alternative approaches, like those represented by neurocontrollers, which are model-free and based on learning.

Self-Assessment Questions

1. What is the size of the actuator space, the configuration space, and the task space, respectively, in figure 6.1? What is it in the case of our soft arm in figure 1.1(a), with the four types of actuations analyzed in chapter 4 and reported in figure 6.1?

2. Apply the controllers of figures 6.2 and 2.13 to the case of the soft arm in figure 1.1(a) with cable actuation.

3. Set the size and the contents of the input and output layer of a neural network controlling the robot of question 2.

4. Neurocontrollers are generally less accurate than model-based controllers. What do you think is the advantage of using them with soft robots?

5. Is it computationally more efficient to use model-based controllers or neurocontrollers?

Further Readings

On Neural Networks

Aggarwal, Charu C. 2023. *Neural Networks and Deep Learning: A Textbook*, 2nd ed. Cham, Switzerland: Springer. https://doi.org/10.1007/978-3-031-29642-0.

Mordechai, Ben-Ari, and Francesco Mondada. 2018. *Elements of Robotics*. Cham, Switzerland: Springer.

On Soft Robot Control

George Thuruthel, Thomas, Yasmin Ansari, Egidio Falotico, and Cecilia Laschi. 2018. "Control Strategies for Soft Robotic Manipulators: A Survey." *Soft Robotics* 5 (2): 149–163. https://doi.org/10.1089/soro.2017.0007.

7 Soft Robotics in Practice

Chapter Objectives

- To put the lessons learned from previous chapters together
- To see how previously learned concepts and techniques can be put in practice
- To integrate previously learned components into one soft robotic system

7.1 Overview

Although we encourage an integrated view where materials merge with actuators and sensors and the body mechanics perform a part of control, the previous chapters have so far covered only what we consider the main components of a soft robot. That was done for educational purposes, and hopefully we achieved the purpose. Let us now challenge ourselves in gathering that wealth of knowledge around a specific case study as an exercise to put all jigsaw pieces together to come up with a soft robot.

The case we are going to analyze is one of the first and most extreme models for soft robotics: an octopus. It is an ideal example since it is completely soft, with no bones or cartilage, so we are not puzzled by the role of soft versus rigid parts. Materials, actuators, and sensors for reaching a similar dexterity are an extremely interesting challenge and test bench for diverse technologies. More than that, the octopus's ability to stiffen its arms and create a sort of deformable skeleton stretches our current technological thinking. Being a mollusk, the octopus shows an unexpectedly rich sensory-motor behavior that supports the idea of embodied intelligence, since its special morphology and dexterity allow for a rich interaction with the environment. An octopus is a benthic animal, living underwater but in contact with the seafloor, which gives us models for fluid and solid interactions. Its control system is highly distributed in the brain and in its eight arms and shows a number of simplifying principles for robotics. Can you find a better model for soft robotics?

7.2 Bioinspired Principles

According to the method described in chapter 3 (outlined in figure 3.3), we start with the extraction principles from the observation of the biological system. An octopus is a goldmine of principles for a soft roboticist. We focus on three main functions that are relevant to robotics: grasping, walking, and swimming. But first, let us anchor on a few terms related to the octopus's basic anatomy and neurophysiology.[1]

7.2.1   Octopus Basics

An octopus has eight arms that are identical (except for a reproduction-related appendix in one of them, which is not relevant to our robotics purposes). The octopus arms can bend in any direction and at any point along the arm length. They can elongate and shorten, and most interestingly, they can stiffen. In fact, they are composed of soft tissues, mostly muscles, that are arranged in a special way. Similar arrangements are only found in the arms and tentacles of other cephalopods (squids and cuttlefish, to give a few examples), in the elephant trunk, and in tongues of reptiles and mammals (including our own tongue). These muscular structures are named muscular hydrostats as their volume keeps constant during contractions.

In the octopus arm, four main longitudinal muscles are equally spaced along the arm length, transverse muscles are arranged radially in the cross-section all along the arm, and oblique muscles are wrapped around the arm, both clockwise and counterclockwise; see figure 7.1(c). Broadly speaking, longitudinal contractions reduce the length of the respective side of the arm, generating a bending. Contractions of all longitudinal muscles at the same time, with the other muscles relaxed, reduce the arm length. Contractions of the transverse muscles reduce the diameter of the respective cross-section instead. Contractions of all transverse muscles at the same time, with the other muscles relaxed, increase the arm length. Such elongation reaches almost 80 percent of length at rest. However, it corresponds to around 20 percent of active contraction of transverse muscles, while the longitudinal muscles are passive in fact.

The oblique muscles have the specific role of generating arm torsions, in both directions. You may recall from chapter 4 that there is a magic angle of $54.7°$ in woven structures where the side effect of contraction or extension is nullified. The angle of oblique muscles matches it exactly so that their function is fully decoupled from shortening and elongation.

When the arm muscles are contracted at the same time, the overall stiffness is increased, with no shape change. In other words, the combination of muscle contractions can generate a desired posture, and then the arm can be stiffened in that posture. It is like creating a skeleton with rigid links and joints that can change over time.

Octopus arms have one or two rows of suckers on the ventral surface,[2] as outlined in figure 7.1(a). Their anatomy includes a cavity, and adhesion is ensured by creating a negative pressure inside it, through contractions of specific sucker muscles. In addition to their adhesion role, the suckers are highly sensitive to mechanical and chemical stimuli. The octopus skin is also densely sensorized for mechanical stimuli. It is composed of a highly stretchable tissue and can change color and texture for camouflage.

The octopus nervous system is the most complex nervous system of all invertebrates, and it is greatly distributed. The very central part of each octopus arm is a longitudinal canal that contains nervous fibers. The nerve cords distributed in the eight arms that together contain more neurons (around 350 million) than the brain itself (around 150 million neurons). Most notably, they control a good part of the arm movements locally. A detached arm can still produce effective grasping movements by local stimulation, with

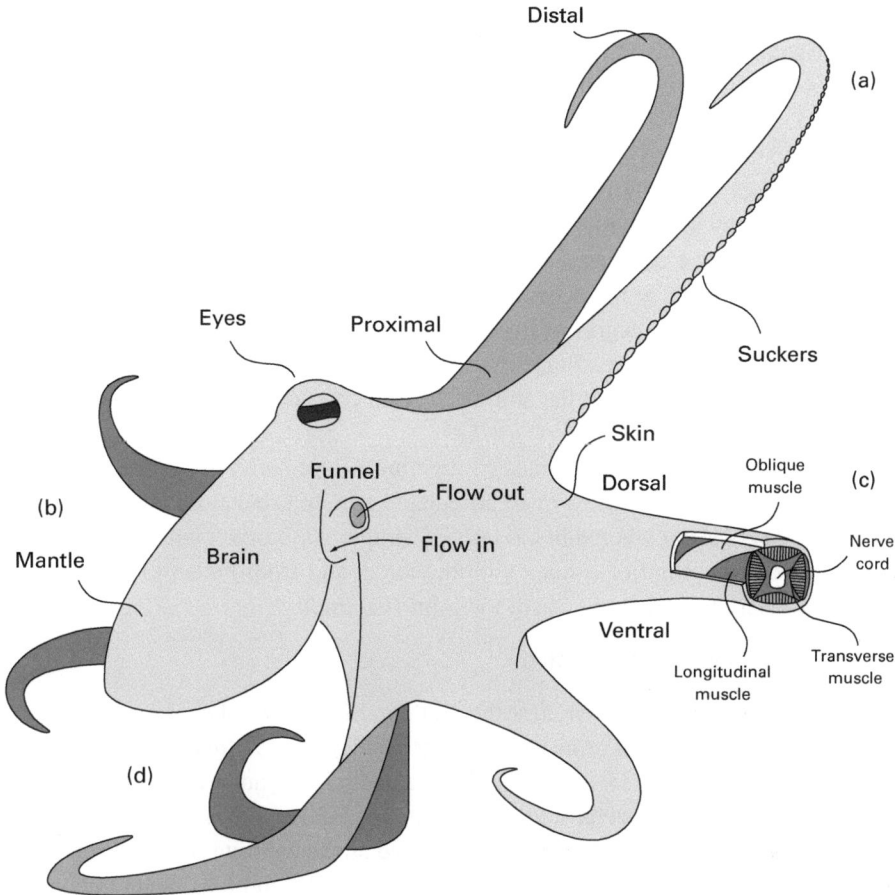

Figure 7.1
A view of the octopus anatomical elements and terms cited in the chapter, including a depiction of the arm muscular hydrostat, labeled (c).

no brain control. On an anatomical note, the nervous fibers inside the arm show a coiled arrangement that complies with arm elongations.

The octopus brain is contained in the mantle—figure 7.1(b)—together with other organs. A big part of the brain is devoted to vision, with two optic lobes with almost 60 million neurons each. The octopus's vision is well developed, and the eye itself has a high number of receptors (20 million) compared to other mollusks.

Figure 7.1 recalls some of the terms related to the octopus anatomy, and you may refer to it when learning about the three main functions we are going to study in the following sections.

7.2.2 Reaching and Fetching

The way arms are used by the octopus for reaching and fetching is well characterized in biology. When an octopus identifies a target by vision, it launches two arms in its direction, in a ballistic movement that cannot be corrected before its end. If the target is missed, the reaching movement is started over. The two arms reach out for the target with a bending wave that propagates from the arm base toward its tip; see figure 7.1(a). The arm segment before the bend is stiffened while the distal segment after the bend is passive. When the target is touched, the passive distal part tends to wrap around it by inertia.

This reaching strategy has several advantages in robotics. First of all, the bend reduces the area of the arm surface opposing resistance to the water, thus minimizing water drag. The bending is accentuated by the water itself as an external mechanical action. Then, the bending propagation is obtained by a sequential activation of motoneurons that is controlled locally. The control effort is reduced by the activation being sequential. Despite the many muscles involved and a virtually infinite degree of freedom (DOF) to control, the brain only sets three control parameters. The octopus reaching strategy is an example of embodied intelligence simplifying control by accounting for the arm mechanical properties and its interaction with water.

Fetching a target to the mouth is obtained by creating a joint between two stiffened arm segments. You may see it as the elbow of an articulated arm, to use a more familiar image. In control terms, this is also obtained in a relatively simple way: two sequential stiffening activations start from the point of contact with the target and from the arm base, respectively. The point where they meet is where the joint is created.

7.2.3   Underwater Legged Locomotion

Octopuses are benthic animals that live near the seafloor. Sometimes, they show a walking-like locomotion that makes use of arms as legs. The strategy is again quite smart in taking advantage of the water environment. Octopuses have a neutral buoyancy, and they raise their body off the ground, so they do not have to compensate for friction like crawling animals do. They use only two of the eight arms, and precisely the two on the opposite side of the intended locomotion direction. The movement of these two arms is cyclic and thus easier to control: each arm is shortened, attached to the seafloor, elongated and stiffened to push the body forward, and then detached and shortened again. This walking pattern based on shortening/elongating along the arm longitudinal axis reduces the water drag that is instead dramatic when displacing rigid legs for walking in water.

7.2.4   Pulsed-Jet Swimming

Octopuses can swim for fast and short escapes from predators. They use their mantle and funnel—for reference, see figure 7.1(b)—for generating jet propulsion. They fill the mantle with water and expel it through the funnel with a mantle contraction. They can repeat this sequence a few times and obtain a pulsed-jet propulsion. To inflate the mantle, they reduce its wall thickness in order to increase its internal volume and create the negative pressure that makes water flow in. To expel water, they contract the muscles that squeeze the mantle.

The control is again relatively simple since it consists of one signal for contracting the muscles involved at the same time, both for filling the mantle and for expelling the water. The morphology of the mantle and the funnel and the frequency of contractions produce the correct fluid dynamics effect for optimal propulsion (i.e., ring vortices out of the funnel).

You may recall the basic idea of embodied intelligence in this and other octopus strategies. The system and the interaction with the environment are complex, but they contribute to control, which is simplified.

7.3 Materials, Actuators, Sensors

7.3.1 Materials

One of the first design questions is whether any material available for robotics matches the mechanical properties of the octopus arm tissue. A similar characterization as the one described in chapter 4 can be done on a sample of the biological tissue, in compression and tension. A stress-strain curve would show a hyperelastic behavior, meaning that for increasing stretch, the stress increases in a nonlinear way. The same happens in some commercial silicones, like Dragon Skin produced by Smooth-On, Inc. In terms of density, the octopus arm tissue is quite similar to seawater (i.e., 1.02 kg/m^3 for seawater and 1.04 kg/m^3 for the octopus arm). Commercial silicones with similar stress-strain curve also have a similar density of 1.04 kg/m^3. Silicone can be cast in diverse morphologies, including one similar to the octopus arm tapered rod, with decreasing cross-sectional diameter from base to tip.

7.3.2 Tendon-Driven Actuation

A first idea for actuating a silicone conical rod is embedding cables inside the material, as seen in chapter 4 (section 4.3.1). We learned that an octopus arm has four longitudinal muscles; we may think of anchoring four cables to the arm tip, have them run longitudinally, equally spaced, and pulled from external motors at the base. Single activations of one cable would produce a shortening on one side and then a bending on that side. Pulling the four cables at the same time would instead shorten the arm overall. This is already a good start, but we miss the possibility of a sequential activation, base to tip, that is key for the reaching movement. An improvement of this tendon-driven design is adding cables anchored at shorter distances from the arm base. Their sequential activation would then be possible with separate motors. However, the reaching movement is based on stiffening the proximal part of the arm before the bending. Then, we need transverse muscles, too. Still building on the previous tendon-driven solution and considering that longitudinal and transverse muscles are activated at the same time for stiffening, we can couple them. A same cable can run longitudinally from the base to a pulley-like element close to the midline, from which it goes transversally to anchor itself on the arm external surface. See figure 7.2(a) for a pictorial view of the arrangement described.

Section 7.2.4 explains how a relatively simple mantle contraction allows the octopus to swim with a pulsed-jet propulsion, provided the morphology and mechanical properties of the mantle and the funnel and the contraction frequency are appropriate. A mantle-like shell and a funnel-like aperture can be fabricated by casting silicone. A tendon-driven approach can be used to actuate the mantle and contract it to expel water through the funnel. Cables anchored to the mantle walls reduce its diameter when pulled by one motor; see figure 7.2(c).

A slightly different design of the tendon-driven actuation can be used to implement the cyclic arm movement for walking (refer to section 7.2.3 for the octopus walking principle). A flexible rod embedded in the arm, along the midline, generates the shortening and elongation phases when pulled and pushed by a motor through a simple crank mechanism; see figure 7.2(e). The combination of the arm morphology and the mechanical properties of the material provides appropriate stiffening when interacting with the ground.

7.3.3 EAP as Artificial Longitudinal and Transverse Muscles

To decouple longitudinal and transverse contractions—for instance, for arm elongation/shortening—we may consider smart materials that play the role of artificial longitudinal and transverse muscles. Section 4.3.3 describes EAP as contractile devices that could fit our purpose. Figure 4.11 already shows how a stack of EAP embedded into a soft arm can contract one side and reduce its length, working as a longitudinal muscle. We may think of a design that integrates four longitudinal muscles with a number of transverse muscles that reduce the cross-sectional diameter. Figure 7.2(d) shows this concept.

An EAP can be fabricated with a commercial silicone (section 7.3.1) molded in a thin layer. To keep thickness and stiffness small, electrodes can be produced by sputtering a metal (e.g., gold on the two sides of the silicone film). The film can then be cut into a strip to fold into a stack. We may take the 20 percent transverse contraction of the octopus as a reference for setting our design parameters. With a film thickness of 300×10^{-6} m (μm), a 20 percent contraction means reaching a thickness of 240×10^{-6} m; if the starting thickness is 200×10^{-6} m, then the final one is 160×10^{-6} m. Similarly, with 150×10^{-6} m as initial thickness, we aim at 120×10^{-6} m. Considering the material properties, the stress needed is around 15000 Pa. The force F and the stress S that our EAP generates depend on the geometrical features (area A and thickness t of the EAP), the material properties ε (material dielectric constant and relative permittivity), and the voltage applied V, as expressed in equation (4.24) and equation (4.25), respectively. In the first case of initial $300\mu m$ thickness, with a voltage V of $2000V$, the stress S only reaches slightly above 1000 Pa, and increasing the voltage to $2500V$ and to $2750V$ brings us to 2000 Pa and 3000 Pa, respectively. With a stack area A of 6×6 mm, the latter S corresponds to a force of 0.075 N and to a contraction of about 3.5 percent. According to equation (4.24) and equation (4.25), we may increase the voltage V and reduce the thickness t to attain higher stress values and corresponding contractions. With $t = 200 \times 10^{-6}$ m and $V = 2750V$, S reaches 7500 Pa and $F = 0.18$ N, which corresponds to a 9 percent contraction. We can finally reach 20 percent contraction with the same $V = 2750V$ voltage and $t = 150 \times 10^{-6}$ m.

This design is conceptually correct and theoretically feasible for implementing octopus-like longitudinal and transverse muscles, in a similar arrangement to a muscular hydrostat. Practical considerations for the fabrication of EAPs and their integration into the target arrangement would add important elements to the final performance.

7.3.4 SMA Springs as Artificial Longitudinal and Transverse Muscles

Section 4.3.4 explains how shape-memory alloys (SMAs) like NiTiNOL change shape when heated. A spring arrangement of NiTiNOL wire would generate a contraction along the longitudinal direction if properly set to a contracted initial shape. We may then design a combined longitudinal and transverse arrangement of SMA springs, similar to a muscular hydrostat. The radial SMA springs positioned in the same cross-section can contract at the same time, controlled by one input, and reduce the cross-sectional diameter. Each longitudinal SMA spring can reduce the length of the segment it covers.

Chapter 4 reports a few equations—for instance, equations (4.26) and (4.27)—that describe the many parameters contributing to the SMA strain and the relation with the electrical power injected in the system, without taking into account the material the SMA springs may be embedded in. For reference, see section 4.3.4. Despite their stroke being high enough for our target of 20 percent contraction, the force produced may not be enough to squeeze the material. We then consider the springs as anchored to the external surface of the octopus-like arm, with empty space inside. In fact, in the animal, the connective tissue plays an important role in the overall arm movement, and a woven sleeve inspired to it stands as a plausible arm structure. Chapter 4 discusses the woven sheaths used in McKibben actuators. You may recall again the 54.7° angle that nullifies elongation/shortening, which appears in the octopus arm connective tissue and can be used in our SMA-based arm.

A few design considerations are needed for positioning the transverse springs. A central anchor is needed, close to the midline. Although the octopus arm shows a macro-arrangement of four transverse arcs, as shown in figure 7.1(c), many transverse muscle fibers run radially through the longitudinal fibers so that a radial arrangement of the SMA springs is a good approximation. Their number should be high enough to ensure a uniform diameter reduction. The woven sleeve helps transmit the punctual deformation to the rest of the structure, both transversally and longitudinally, so that we can discretize the number of transverse actuated sections.

Concerning the longitudinal SMA springs, we may consider local contractions and sequential activation and place longitudinal springs in between two actuated cross-sections. We would not achieve an overall (global) bending as would be possible if we used a longitudinal actuator from base to tip. Integrating cables as in section 7.3.2 would be a feasible solution to this purpose. Figure 7.2(b) summarizes the design described. Please refer to the second step of the biorobotics method in figure 3.3, implementing a biomimetic robot for validating the principles.

7.4   Modeling Internal and External Interactions

Our octopus arm has an elongated morphology that can be described by rod models. Length is larger than cross-sectional dimensions, and volume is constant. We may use the general Cosserat's method to describe the arm and its bending. According to equation (5.5), we describe the arm mechanics related to linear and angular velocities with $\mathcal{N}$ terms that assume the following form:

$$\mathcal{N}_{sb,v} = \frac{1}{\rho A}\left(\frac{\partial n}{\partial s} + \bar{n}\right) \tag{7.1}$$

$$\mathcal{N}_{sb,\omega} = \frac{I^{-1}}{\rho}\left(-\omega \times (\rho I \omega) + \frac{\partial c}{\partial s} + \frac{\partial r}{\partial s} \times n + \bar{c}\right) \tag{7.2}$$

where ρ is the density of the medium, I is the moment of inertia, A is the area of the cross-section, r is the position vector of $O(s)$, and n and c are the linear and angular cross-sectional stresses along the arm, respectively. Actuators are added to the model as active internal wrenches. In the case of tendon-driven actuation (section 7.3.2), cable tensions are the variables in play.

Interestingly, the same Cosserat's approach can be used with shell morphologies. Instead of the rod cross-sections, the micro-solids here are rigid fibers, transversally attached to

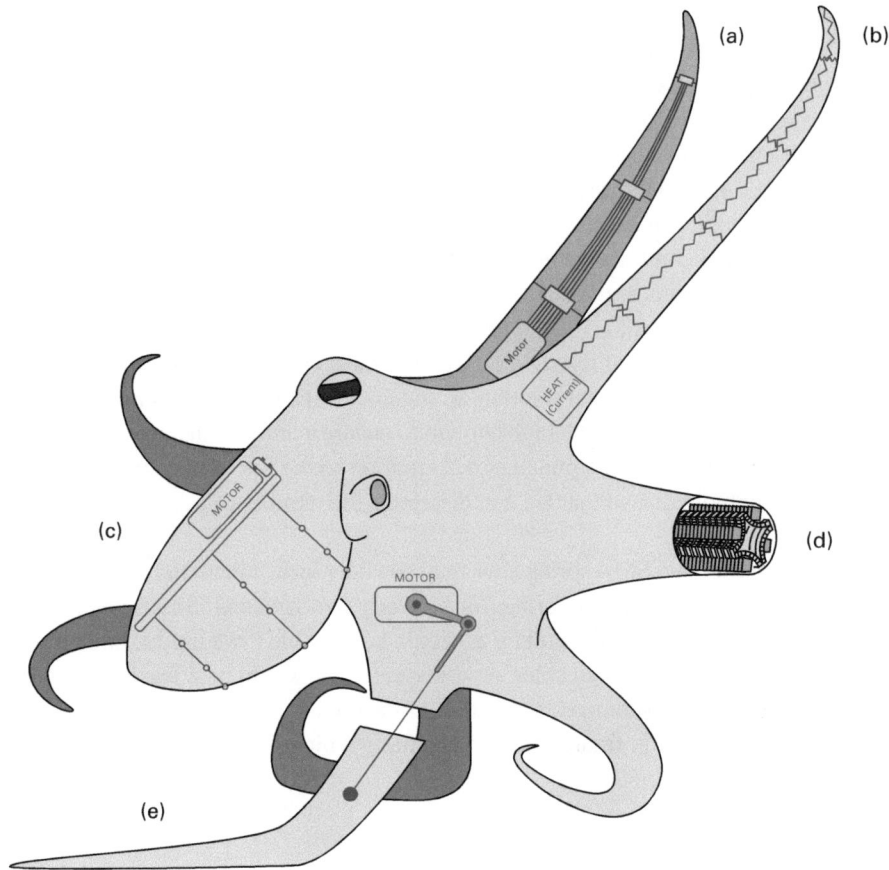

Figure 7.2
Pictorial view of the different actuation technologies that can be used for an octopus-like soft robot.

the shell mid-surface. Very similar equations describe their deformations. Shells are interesting in our octopus case study because they decently approximate the morphology of the octopus mantle. A Cosserat's shell model can describe the deformations of the pulsed-jet swimmer discussed in section 7.3.2, under the action of the tendon-driven actuation integrated in the model as internal wrenches.

The models of internal interactions described above need to be complemented with the models of external interactions, as discussed in chapter 5. The complex fluid dynamics involved in the pulsed-jet swimming can be modeled with lumped parameters, as explained in section 5.4.2, by accounting for the force contributions from added mass, drag, lift, and buoyancy. Coupling such a lumped parameter model with the shell model gives a complete description of the swimming behavior.

Modeling underwater walking (described in sections 7.2.3 and 7.3.2) requires modeling the interactions with the solid ground as well as the fluid underwater environment. A typical model for legged locomotion, used in robotics, is the spring-loaded inverted pendulum (SLIP), where the whole body is modeled as a point mass m, attached to a massless leg with linear spring k, forming an angle of attack θ with the ground. The underwater-SLIP (U-SLIP) adds key parameters from the underwater environment—namely, added mass, drag, lift, and buoyancy. Among the differences in the resulting locomotion pattern are an opposite angle of attack and longer flight phases. Such a model is a step forward in the biorobotics method of figure 3.3, abstracting the principle for underwater legged locomotion.

7.5   Model-Based and Learning-Based Control

When a model of internal interactions is available, implementation of model-based control is possible. Considering the tendon-driven actuation of section 7.3.2, with the Cosserat's model, we have the relation between the cable tensions (contained in a vector $\mathbf{T}$) and the arm deformation, which correspond to a tip position $\mathbf{x}$. Such relations can be inverted, and cable tension vector $\mathbf{T}$ can be derived for a desired $\mathbf{x}_d$, as discussed in chapter 6. Analytical methods are not always usable because of the ordinary differential equations involved, while iterative methods can find suitable solutions. The Jacobian method, an iterative algorithm based on the Jacobian, finds $\mathbf{T}$ starting from the desired tip position $\mathbf{x}_d$. A simple controller can ensure that the motor attached to a cable makes it reach the target tension.

The same job can be done by a neural network, as a neurocontroller that finds $\mathbf{T}$ for a given $\mathbf{x}_d$. The network input layer would encode $\mathbf{x}_d$, so it needs six neurons for a full 3D description. The network output layer would encode $\mathbf{T}$, so it needs as many neurons as the number of cables. In between, we may think of a single layer, fully connected to the input layer, on one side, and to the output layer, on the other side. The supervised learning phase requires a training set of correct input-output pairs, used with the backpropagation technique explained in section 6.4.

Take a moment and examine what the differences in the two approaches are and which of the two would show better performance and under what conditions. In terms of computational time, the Jacobian method is more expensive, since new calculations need to be done for each new input. Once the neural network is trained, a neurocontroller gives an output in a relatively short time, since the computation involved in neuron activations is faster. However, the learning time should also be put in the count. In terms of accuracy, neural networks approximate solutions, which are not as accurate as those obtained with model-based methods like the Jacobian method here. However, the accuracy of model-based methods depends on the accuracy of the models they use. A soft robot like the arm of our example, cast in silicone and manually embedded with cables, is not as accurate as its theoretical model. For this reason, a model-based method may be inaccurate and have variable performance from one arm to another, even if fabricated with the same process. The unavoidable differences in the physical arms generate variable inaccuracies in model-based methods. If the neurocontroller is trained on the physical arm instead, its performance tends to be the same, for different arms.

7.6   Summary

This chapter recalled the main concepts and some of the techniques of previous chapters around the case study of an octopus-like robot. From chapter 3, we take the method for identifying the principles in the octopus anatomy, neurophysiology, and behavior and adopt it in the design of the octopus-like robot. From chapter 4, we take some considerations about soft materials and a few technologies for actuating the octopus-like arm and mantle. From chapter 5, we use the Cosserat's method to model the arm and mantle deformations and the lumped parameter models for interaction with the water and with the seafloor, combining them together for pulsed-jet swimming. From chapter 6, we take the two possible approaches to control, based on models and based on learning, to draw some comparative discussion. Figure 7.2 graphically summarizes the design discussed in the chapter.

Self-Assessment Questions

1. Which sensing technologies can be used in the octopus-like arm?
2. What effect does the water environment have on the octopus-like robot?
3. Elaborate on the pros and cons of model-based versus learning-based controllers.
4. Propose an application for the octopus-like arm.
5. Propose your own case study for the soft robotics technologies learned throughout the book.

Further Readings

On Cosserat's Models of Octopus-like Robot

Renda, Federico, Matteo Cianchetti, Michele Giorelli, Andrea Arienti, and Cecilia Laschi. 2012. "A 3D Steady-State Model of a Tendon-Driven Continuum Soft Manipulator Inspired by the Octopus Arm." *Bioinspiration and Biomimetics* 7 (2): 025006. https://doi.org/10.1088/1748-3182/7/2/025006.

On the Octopus

Hochner, Binyamin, Letizia Zullo, Tal Shomrat, Guy Levy, and Nir Nesher. 2023. "Embodied Mechanisms of Motor Control in the Octopus." *Current Biology* 33 (20): R1119–R1125. https://doi.org/10.1016/j.cub.2023.09.008.

On Octopus Bioinspiration

Margheri, Laura, Cecilia Laschi, and Barbara Mazzolai. 2012. "Soft Robotic Arm Inspired by the Octopus: I. From Biological Functions to Artificial Requirements." *Bioinspiration and Biomimetics* 7 (2): 025004. https://doi.org/10.1088/1748-3182/7/2/025004.

Mazzolai, Barbara, Laura Margheri, Matteo Cianchetti, Paolo Dario, and Cecilia Laschi. 2012. "Soft-Robotic Arm Inspired by the Octopus: II. From Artificial Requirements to Innovative Technological Solutions." *Bioinspiration and Biomimetics* 7 (2): 025005. https://doi.org/10.1088/1748-3182/7/2/025005.

On Octopus-like Robot with EAP Actuation

Cianchetti, Matteo, Virgilio Mattoli, Barbara Mazzolai, Cecilia Laschi, and Paolo Dario. 2009. "A New Design Methodology of Electrostrictive Actuators for Bio-Inspired Robotics." *Sensors and Actuators B: Chemical* 142 (1): 288–297. https://doi.org/10.1016/j.snb.2009.08.039.

On Octopus-like Robot with SMA Actuation

Laschi, Cecilia, Matteo Cianchetti, Barbara Mazzolai, Laura Margheri, Maurizio Follador, and Paolo Dario. 2012. "Soft Robot Arm Inspired by the Octopus." *Advanced Robotics* 26 (7): 709–727. https://doi.org/10.1163/156855312X626343.

On Octopus-like Robot with Tendon-Driven Actuation

Cianchetti, Matteo, Andrea Arienti, Maurizio Follador, Barbara Mazzolai, Paolo Dario, and C. Laschi. 2011. "Design Concept and Validation of a Robotic Arm Inspired by the Octopus." *Materials Science and Engineering: C* 31 (6): 1230–1239. https://doi.org/10.1016/j.msec.2010.12.004.

On Model-Based and Learning-Based Control of an Octopus-inspired Arm

Giorelli, Michele, Federico Renda, Marcello Calisti, Andrea Arienti, Gabriele Ferri, and Cecilia Laschi. 2015. "Neural Network and Jacobian Method for Solving the Inverse Statics of a Cable-Driven Soft Arm with Non-constant Curvature." *IEEE Transactions on Robotics* 31 (4): 823–834. https://doi.org/10.1109/TRO.2015.2428511.

On Pulsed-Jet Swimming Modeling

Renda, Federico, Francesco Giorgio Serchi, Frédéric Boyer, and Cecilia Laschi. 2015. "Structural Dynamics of a Pulsed-Jet Propulsion System for Underwater Soft Robots." *International Journal of Advanced Robotic Systems* 12 (6): 68. https://doi.org/10.5772/60143.

On the U-SLIP Model

Calisti, Marcello, and Cecilia Laschi. 2017. "Morphological and Control Criteria for Self-Stable Underwater Hopping." *Bioinspiration and Biomimetics* 13 (1): 016001. https://doi.org/10.1088/1748-3190/aa90f6.

8 Conclusions

Chapter Objectives

· To take an overall look at the full picture of soft robotics, as depicted in this textbook

· To grow a critical thinking on soft robotics technologies and the abilities they enable in robots

· To learn about the possible applications of soft robots

· To build a vision for the progress of robotics

8.1 Overview

This chapter takes a step back and looks at the overall picture of what we have learned about soft robotics. We discuss what soft robots can do today and which applications they can respond to, today and tomorrow. Then we peek into the future to see how soft robotics can help shape the next generation of robots.

8.2 Our Journey in Soft Robotics

We are reaching the end of our journey, a deep dive in soft robotics. We strolled around its underlying principles and stopped by the technologies that enable building soft robots, with their soft materials, deformable structures, smart actuators, and soft sensors. We wandered along the techniques for modeling soft robots and the way they interact with their surroundings, and we visited the world of soft robot control through model-based and learning-based techniques. Before ending our journey, we gathered our wealth of knowledge around a specific case study, the octopus, to see a soft robot in practice. What is next?

Soft robotics is a young and growing fast field, so more findings will be coming soon after this book is published. The basic principles, approaches, and techniques learned here are going to stand as a solid base on which to build up state-of-the-art technological knowledge and experiences.

8.3   Soft Robot Abilities

Soft robots will be able of more and more functions, from classical yet revisited manipulation and locomotion to fancier growing, morphing, self-healing, biodegrading. Some of them are explored and implemented already, at least in research labs, but many more are there, just waiting to be discovered by researchers' creativity.

8.3.1   Soft Robot Manipulation

A key ability in robots, manipulation has been the focus of robotics research and development since early times. In industrial robotics, specialized grippers are sometimes used for different tasks. However, robotic hands, with humanlike universal function, have been the goal for decades. Fingers that adapt to object shapes simplify control, according to the principle of embodied intelligence. Soft fingers represent an effective approach in this respect. Others have reviewed the approaches and the technologies for building soft grippers (Shintake et al. 2018).

An alternative to finger-based grippers is the so-called universal gripper, built on the principle for granular jamming explained in section 4.4. A simple round-shaped elastic membrane is filled with grinded coffee and connected to a vacuum pump (Brown et al. 2010). When put in contact with an object, it can conform to its shape. When air is removed and the gripper stiffens, the object is held in a stable grasp. A very wide range of object shapes can be grasped with this solution.

8.3.2   Soft Robot Locomotion

In nature, locomotion assumes a variety of forms. Legged locomotion is well-known and widely adopted in robotics, providing advantages over mobility on wheels, in unstructured environments. Biped locomotion of humanoid robots is well described by robust models that are used for designing and for controlling robot walking. Multi-legged robots have also been built successfully. From previous discussions about embodied intelligence, we see that walking benefits from compliant joints and soft leg/feet parts, especially for simplifying control by mechanically compensating for terrain unevenness. One of the widely adopted models for legged locomotion, spring-loaded inverted pendulum (SLIP), models a leg as a spring (Or and Moravia 2016). Moving to more dynamic locomotion patterns, like running, compliance becomes more important, or better, necessary. It is knee compliance that allows humanoid robots to run.

A soft body enables forms of locomotion other than walking or running. Body deformations themselves produce limbless locomotion patterns that resemble snakelike to wormlike crawling. In the case of snakelike crawling, despite a backbone that is often built as a hyper-redundant chain of short rigid links, soft materials improve flexibility and the anisotropic friction that is at the base of locomotion. Wormlike and caterpillar-like locomotion is based on a body bending on the vertical plane. Higher friction on the front contact point brings the body close to it. Shifting high friction to the back contact point and straightening the body produces a forward movement.

Peristalsis is widely observable in nature as a mechanism for locomotion and for transportation. It is based on tubelike structures that reduce their diameter at a given point and transfer this diameter reduction forward, in a wavelike movement that generates motion in one direction. In the gastrointestinal tract, peristalsis generates one-direction transportation. In wormlike robots, it generates locomotion.

Jumping is a fast form of locomotion enabled by elastic elements that store energy and release it at once, generating a ballistic flight. In swimming as well, despite its diverse forms, common tracts are that flexible bodies are needed and elastic structures that store energy and release it to produce propulsion in water.

In all of these cases, having a soft body improves locomotion, or it is just necessary to achieve it. Fundamental models with mathematical formulations and descriptions of the parameters that come into play are described elsewhere (Calisti, Picardi, and Laschi 2017). Soft robotics technologies are the way to implement them and provide robots with forms of locomotion that were not possible before.

8.3.3 Growing Robots

A striking difference between natural and artificial systems is that living organisms grow during their life. Growth is instrumental to many purposes in nature, and in some species it is also a form of movement. This is the case with plants, which actually move by growing. Their highly sensorized roots move underground in search of nutrients. In climbing plants, shoots grow in search of support and light. Inspired by those plant behaviors, growing robots are developed by increasing their length by adding material. Seminal work has been done by Barbara Mazzolai at the Italian Institute of Technology, as mentioned in section 3.4.2. A recent robot inspired by climbing plants, growing through an embedded additive manufacturing mechanism, is described in Del Dottore et al. (2024).

8.3.4 Self-Healing Robots

Materials exist that can recover their properties after being damaged. Some elastomers can recover their mechanical characteristics by a light or thermal trigger after being punctured or cut. This self-healing property can be used for building soft robots that go back to working after sustaining damage. An example of a self-healing soft hand is described in Terryn et al. (2017). Its fingers work with the pneumatic actuation described in section 4.3.2. When cut, air leakage prevents them from properly bending. By self-healing, the gripper recovers full performance.

8.3.5 Biodegradable Robots

In soft robotics, we are rethinking the materials that robots are built of. The focus is on their mechanical properties, but we may add an additional requirement to our design and look for materials that are also biodegradable. The main purpose is building robots with minimal impact at the end of their lifespan (see also section 8.4). However, degradability can be used in a functional way during the robot's lifetime. Think, for example, of robots used underwater, where degradable elements may let robot parts detach and resurface at a given time. We need materials with degradability that can be adapted to the robot's working conditions and that can be programmed. Materials exist with such characteristics, and others are synthesized in labs to match this goal.

8.3.6 Biohybrids

Biohybrid is intended here to mean the combination of artificial and biological components. Biological cells or tissues can be integrated into artificial systems and become biohybrid robots. For instance, cardiomyocytes have been demonstrated to work as actuators in hybrid miniature robots, but more solutions are being investigated (Ricotti et al. 2017). Artificial intelligence (AI) methods can help find suitable morphologies, as in the case of Xenobots, built by integrating biological tissue on computer-generated designs tested in simulation (Kriegman et al. 2020). A recent book (Raman 2021) describes biofabrication, or the technology that enables building with biology, together with its use in tissue engineering, organs-on-a-chip, lab-grown meat and leather, and biohybrid machines.

8.4   Soft Robot Applications

Soft robots are already finding applications in diverse fields, from medicine to personal assistance, from agriculture to food industry, from exploration of the abyss to outer space. We list a few cases in the following sections. These cases give just a partial view of the soft robotics field and may be outdated by the time you are reading this book. Hopefully, the soft robotics field will keep growing, and soft roboticists will keep exploring new ideas at the pace we are used to.

8.4.1   Biomedical Applications of Soft Robots

In the biomedical field, soft robots are finding a diversity of applications, inside and outside the human body.

Inside the human body, soft robotics technologies can be used to build endoscopes that softly move in tiny spaces and around organs and that can increase their stiffness to do operations such as taking images and sampling tissues. This application case provides advantages from the clinical viewpoint, since the higher dexterity of soft endoscopes allows for enlarging the reachable area.

Outside the human body, soft robots can assist disabled and elderly people for their own independence or for supporting caregivers. Soft robot arms can safely operate in close proximity to the patient, offer support for daily tasks, and become a tool for human caregivers. They enable scenarios where the robot takes over the physical effort while caregivers retain the role that requires their professional skills and human touch.

Soft robots may support rehabilitation, with soft arms promoting physical therapy with appropriate stiffness profiles. Being soft facilitates physical contact with human patients to help them perform prescribed exercises. A soft robot can support a limb movement by also adjusting its stiffness to the patient's motor ability and recovery phases.

Soft robotics technologies are employed in wearable robots that can either promote rehabilitation or assist in the activities of daily living. Soft materials and textiles, coupled with soft actuators, are better suited for being worn. They promote scenarios where wearable robots are comfortable enough to be used more widely, beyond the limited context of a therapy session. They can help improve the user's motor function to the level required for daily tasks.

Artificial organs, either as replacements or for medical studies, can also benefit from the materials and technologies that are used in soft robotics to improve realistic appearance and motion. The replica of a given organ can be built with soft materials of proper stiffness, coupled with actuators that generate contractions or with smart materials. Such replicas would be a helpful tool for training doctors, for studying pathologies, or even for simulating the effect of treatment and procedures.

8.4.2   Soft Robots in Industry

Soft robotic grippers have industrial applications, such as in the food industry or in agriculture, as adaptable and delicate end effectors for grasping objects of diverse shapes and fragility. Still in industry, the wearable robots mentioned in section 8.4.1 are finding application as support to workers.

8.4.3   Underwater Soft Robots

Underwater soft robots are opening the way to benthic exploration as complementing pelagic navigation of underwater vehicles. The ocean bottom is largely unexplored, and soft robots inspired by benthic animals, as in our case study of chapter 7, stand as novel tools for studying biodiversity and monitoring pollution. Similar purposes can be pursued by soft robots in the wild, putting in place autonomous behavior with energy scavenging and self-healing.

8.4.4   Space Soft Robotics

Humankind is expanding to outer space, and pollution is a reality across our planet, too. Soft robots are being explored for debris collection and other space operations.

8.5   A Vision for Future Soft Robots

The field of soft robotics presents a great opportunity for devising robots that are more sustainable, in a broad sense. Soft robotics paves the way to mend the current gap between nature and technology that uses large portions of the planet's resources, both materials and energy, that produce growing e-waste and raise social and economic concerns. Soft robots can be made of materials that degrade at the end of their lifetime or can be recycled at least; they can be made of materials that can self-heal or multifunctional materials triggered by external, natural stimuli. Using embodied intelligence, they reduce their energy needs and may work with the energy amounts scavenged from the natural environment. Soft robots can develop, learn, and grow, thus adapting themselves to the tasks at hand, in new forms of interaction with human beings. Future soft robots are to be smoothly integrated with the natural environment and the human society.

It is time to end this journey. Take your own step, now, and lend your hand to advancing soft robotics knowledge and putting soft robots to work in support of humanity.

Self-Assessment Questions

1. Be inspired by soft robot design and technologies. What additional abilities may soft robots exhibit?

2. What are the possible advantages in having a soft robot for assistance to an elderly person?

3. Discuss the pros and cons of the use of the actuation technologies learned in chapter 4 in an underwater soft robot.

4. Elaborate on the need for accuracy versus adaptability for soft robots in industry. Which is more important, in which sectors?

5. Propose your own application for a sustainable soft robot.

Further Readings

On Biomedical Soft Robots

Cianchetti, Matteo, Cecilia Laschi, Arianna Menciassi, and Paolo Dario. 2018. "Biomedical Applications of Soft Robotics." *Nature Reviews Materials* 3 (6): 143–153. https://doi.org/10.1038/s41578-018-0022-y.

On Soft Robot Abilities

Laschi, Cecilia, Barbara Mazzolai, and Matteo Cianchetti. 2016. "Soft Robotics: Technologies and Systems Pushing the Boundaries of Robot Abilities." *Science Robotics* 1 (1): eaah3690. https://doi.org/10.1126/scirobotics.aah3690.

On Underwater Soft Robots

Mazzolai, Barbara, Federico Carpi, Koichi Suzumori, Matteo Cianchetti, Thomas Speck, Stoyan K. Smoukov, Ingo Burgert, et al. 2022. "Roadmap on Soft Robotics: Multifunctionality, Adaptability and Growth without Borders." *Multifunctional Materials* 5 (3): 032001. http://iopscience.iop.org/article/10.1088/2399-7532/ac4c95.

On a Vision for Future Soft Robots

Mazzolai, Barbara, and Cecilia Laschi. 2020. "A Vision for Future Bioinspired and Biohybrid Robots." *Science Robotics* 5 (38): eaba6893. https://doi.org/10.1126/scirobotics.aba6893.

On Wearable Soft Robots

Thalman, Carly, and Panagiotis Artemiadis. 2020. "A Review of Soft Wearable Robots That Provide Active Assistance: Trends, Common Actuation Methods, Fabrication, and Applications." *Wearable Technologies* 1: e3. https://doi.org/10.1017/wtc.2020.4.

Notes

Chapter 1

1. Often, 1961 is also used as the birthdate of robotics. The robotics community celebrated 50 years of robotics in the September 2010 issue of the *IEEE Robotics and Automation Magazine* (vol. 17, no. 3).

2. If you are not familiar with robots and robotics, do not worry. We review the basics in chapter 2.

3. Morphology is the form and structure of an organism or any of its parts (Merriam-Webster Dictionary).

4. RoboSoft is a Coordination Action on Soft Robotics that was funded by the European Commission from 2013 to 2016. The RoboSoft community accounted for 34 member institutions for a total of more than 100 scientists.

Chapter 2

1. A robot is an autonomous system that exists in the physical world, can sense its environment, and can act on it to achieve some goals (Matarić 2007).

2. A more correct term is "pose," comprising position and orientation (used in section 1.5.4). However, position is used with the same meaning throughout the book.

3. **K** parameters are now diagonal matrices of coefficients for multiplication with vectors, to account for **u** and **q** being vectors.

4. A formal definition of strain is found in section 4.2.1.

Chapter 3

1. "Overview," PLANTOID Project website, https://plantoidproject.eu/.

Chapter 4

1. See definition 4 (given in chapter 1): soft robots/devices that can actively interact with the environment and can undergo "large" deformations relying on inherent or structural compliance.

2. SI is the International System of Units. We use SI in the rest of the book, based on the MKS unit system (meter-kilogram-second).

3. NiTiNOL is a Ni-Ti alloy, invented in 1963 at the Naval Ordnance Laboratory (NOL); its name stands for nickel, titanium, NOL. It is the most commonly applied shape-memory alloy available in wires, ribbons, sheets, and tubes.

4. It is worth mentioning the use of elastic actuators in robotics. These actuators contain an elastic element that can be inserted either in series (known as a series elastic actuator, or SEA) or in parallel to the other elements (motor and gearbox). The stiffness or impedance of the elastic element can be adjusted in variable stiffness/impedance actuators.

Chapter 5

1. Velocity twist field is a representation of the linear and angular velocity of a rigid body.

2. In screw theory, the concept of a wrench refers to the force and torque vectors that arise in applying Newton's laws to a rigid body.

3. Please note that the first two terms on the right-side of the equation (N_{sb} and C_{int}) can assume any of the forms seen before for the soft body and for the internal interactions, depending on the modeling approach adopted.

Chapter 7

1. Octopuses are found in diverse habitats of different latitude and depth. Many species exist that vary, sometimes greatly, in morphology and size. We mostly refer here to common features of the common octopus (*Octopus vulgaris*) when not otherwise specified.

2. The number depends on the species.

References

Berthoz, A. 2012. *Simplexity: Simplifying Principles for a Complex World*. An Editions Odile Jacob Book. New Haven, Conn.: Yale University Press.

Berthoz, Alain. 2000. *The Brain's Sense of Movement*. Perspectives in Cognitive Neuroscience. Cambridge, Mass: Harvard University Press.

Bishop, Peter J., Scott A. Hocknull, Christofer J. Clemente, John R. Hutchinson, Andrew A. Farke, Belinda R. Beck, Rod S. Barrett, and David G. Lloyd. 2018. "Cancellous Bone and Theropod Dinosaur Locomotion. Part I—an Examination of Cancellous Bone Architecture in the Hindlimb Bones of Theropods." *PeerJ* 6 (October): e5778. https://doi.org/10.7717/peerj.5778.

Brooks, Rodney Allen. 1999. *Cambrian Intelligence: The Early History of the New AI*. Cambridge, Mass: MIT Press.

Brown, Eric, Nicholas Rodenberg, John Amend, Annan Mozeika, Erik Steltz, Mitchell R. Zakin, Hod Lipson, and Heinrich M. Jaeger. 2010. "Universal Robotic Gripper Based on the Jamming of Granular Material." *Proceedings of the National Academy of Sciences* 107 (44): 18809–18814. https://doi.org/10.1073/pnas.1003250107.

Calisti, Marcello, Giacomo Picardi, and Cecilia Laschi. 2017. "Fundamentals of Soft Robot Locomotion." *Journal of the Royal Society Interface* 14 (130): 20170101. https://doi.org/10.1098/rsif.2017.0101.

Cangelosi, Angelo, Josh Bongard, Martin H. Fischer, and Stefano Nolfi. 2015. "Embodied Intelligence." In *Springer Handbook of Computational Intelligence*, edited by Janusz Kacprzyk and Witold Pedrycz, 697–714. Berlin: Springer Berlin Heidelberg. https://doi.org/10.1007/978-3-662-43505-2_37.

Cutkosky, Mark R., and Sangbae Kim. 2009. "Design and Fabrication of Multi-Material Structures for Bioinspired Robots." *Philosophical Transactions of the Royal Society A: Mathematical, Physical and Engineering Sciences* 367 (1894): 1799–1813. https://doi.org/10.1098/rsta.2009.0013.

Del Dottore, Emanuela, Alessio Mondini, Nick Rowe, and Barbara Mazzolai. 2024. "A Growing Soft Robot with Climbing Plant–Inspired Adaptive Behaviors for Navigation in Unstructured Environments." *Science Robotics* 9 (86): eadi5908. https://doi.org/10.1126/scirobotics.adi5908.

Della Santina, Cosimo, Christian Duriez, and Daniela Rus. 2021. "Model Based Control of Soft Robots: A Survey of the State of the Art and Open Challenges." *arXiv:2110.01358 [Cs, Eess]*, October. http://arxiv.org/abs/2110.01358.

Kim, Sangbae, Cecilia Laschi, and Barry Trimmer. 2013. "Soft Robotics: A Bioinspired Evolution in Robotics." *Trends in Biotechnology* 31 (5): 287–294. https://doi.org/10.1016/j.tibtech.2013.03.002.

Kriegman, Sam, Douglas Blackiston, Michael Levin, and Josh Bongard. 2020. "A Scalable Pipeline for Designing Reconfigurable Organisms." *Proceedings of the National Academy of Sciences* 117 (4): 1853–1859. https://doi.org/10.1073/pnas.1910837117.

Laschi, Cecilia, and Matteo Cianchetti. 2014. "Soft Robotics: New Perspectives for Robot Bodyware and Control." *Frontiers in Bioengineering and Biotechnology* 2. https://doi.org/10.3389/fbioe.2014.00003.

Matarić, Maja J. 2007. *The Robotics Primer*. Intelligent Robotics and Autonomous Agents Series. Cambridge, Mass: MIT Press.

Mengaldo, Gianmarco, Federico Renda, Steven L. Brunton, Moritz Bächer, Marcello Calisti, Christian Duriez, Gregory S. Chirikjian, and Cecilia Laschi. 2022. "A Concise Guide to Modelling the Physics of Embodied Intelligence in Soft Robotics." *Nature Reviews Physics* 4: 595–610. https://doi.org/10.1038/s42254-022-00481-z.

Or, Yizhar, and Moti Moravia. 2016. "Analysis of Foot Slippage Effects on an Actuated Spring-Mass Model of Dynamic Legged Locomotion." *International Journal of Advanced Robotic Systems* 13 (2): 69. https://doi.org /10.5772/62687.

Polygerinos, Panagiotis, Nikolaus Correll, Stephen A. Morin, Bobak Mosadegh, Cagdas D. Onal, Kirstin Petersen, Matteo Cianchetti, Michael T. Tolley, and Robert F. Shepherd. 2017. "Soft Robotics: Review of Fluid-Driven Intrinsically Soft Devices; Manufacturing, Sensing, Control, and Applications in Human-Robot Interaction: Review of Fluid-Driven Intrinsically Soft Robots." *Advanced Engineering Materials* 19 (12): 1700016. https://doi.org/10.1002/adem.201700016.

Raman, Ritu. 2021. *Biofabrication*. MIT Press Essential Knowledge Series. Cambridge, Mass: MIT Press.

Ricotti, Leonardo, Barry Trimmer, Adam W. Feinberg, Ritu Raman, Kevin K. Parker, Rashid Bashir, Metin Sitti, Sylvain Martel, Paolo Dario, and Arianna Menciassi. 2017. "Biohybrid Actuators for Robotics: A Review of Devices Actuated by Living Cells." *Science Robotics* 2 (12): eaaq0495. https://doi.org/10.1126/scirobotics.aaq0495.

Rosenblueth, Arturo, Norbert Wiener, and Julian Bigelow. 1943. "Behavior, Purpose and Teleology." *Philosophy of Science* 10 (1): 18–24. https://doi.org/10.1086/286788.

Rus, Daniela, and Michael T. Tolley. 2015. "Design, Fabrication and Control of Soft Robots." *Nature* 521 (7553): 467–475. https://doi.org/10.1038/nature14543.

Schmitt, Otto H. 1969. "Some Interesting and Useful Biomimetic Transforms." In, 1069:197.

Sharpe, Richard Bowdler. 1868. *A Monograph of the Alcedinidae: Or, Family of Kingfishers*. London: Published by the author. https://doi.org/10.5962/bhl.title.69293.

Shintake, Jun, Vito Cacucciolo, Dario Floreano, and Herbert Shea. 2018. "Soft Robotic Grippers." *Advanced Materials* 30 (29): 1707035. https://doi.org/10.1002/adma.201707035.

Snell-Rood, Emilie. 2016. "Interdisciplinarity: Bring Biologists into Biomimetics." *Nature* 529 (7586): 277–278. https://doi.org/10.1038/529277a.

Terryn, Seppe, Joost Brancart, Dirk Lefeber, Guy Van Assche, and Bram Vanderborght. 2017. "Self-Healing Soft Pneumatic Robots." *Science Robotics* 2 (9): eaan4268. https://doi.org/10.1126/scirobotics.aan4268.

Trivedi, Deepak, Christopher D. Rahn, William M. Kier, and Ian D. Walker. 2008. "Soft Robotics: Biological Inspiration, State of the Art, and Future Research." *Applied Bionics and Biomechanics* 5 (3): 99–117. https://doi .org/10.1080/11762320802557865.

Wang, Liyu, and Fumiya Iida. 2015. "Deformation in Soft-Matter Robotics: A Categorization and Quantitative Characterization." *IEEE Robotics and Automation Magazine* 22 (3): 125–139. https://doi.org/10.1109/MRA.2015 .2448277.

Webb, Barbara, and Thomas R. Consi, eds. 2001. *Biorobotics: Methods and Applications*. Menlo Park, Calif.: AAAI Press/MIT Press.

Yang, Guang-Zhong, James Bellingham, Howie Choset, Paolo Dario, Peer Fischer, Toshio Fukuda, Neil Jacobstein, Bradley Nelson, Manuela Veloso, and Jeremy Berg. 2016. "Science for Robotics and Robotics for Science." *Science Robotics* 1 (1): eaal2099. https://doi.org/10.1126/scirobotics.aal2099.

Index

(Bold page numbers point to definitions or first descriptions of the entry term)

Intelligent Robotics and Autonomous Agents
Edited by Ronald C. Arkin

Hwu, Tiffany J., and Jeffrey L. Krichmar, *Neurorobotics*

Cangelosi, Angelo, and Minoru Asada, *Cognitive Robotics*

Grupen, Roderic A., *The Developmental Organization of Robot Behavior*

Laschi, Cecilia, *Soft Robotics*

Publisher contact:
The MIT Press
Massachusetts Institute of Technology
77 Massachusetts Avenue, Cambridge, MA 02139
mitpress.mit.edu

EU Authorised Representative:
Easy Access System Europe, Mustamäe tee 50,
10621 Tallinn, Estonia
gpsr.requests@easproject.com

Printed by Integrated Books International,
United States of America